SKYLARK

Also by Alice O'Keeffe:

On the Up (2019)

ALICE O'KEEFFE
SKYLARK

A NOVEL

CORONET

First published in Great Britain in 2021 by Coronet

An Imprint of Hodder & Stoughton
An Hachette UK company

This paperback edition published in 2022

1

A CIP catalogue record for this title is available from the British Library

Paperback ISBN 9781529303391
eBook ISBN 9781529303414

Typeset in Charter by Palimpsest Book Production Limited,
Falkirk, Stirlingshire
Printed and bound in Great Britain by Clays Ltd, Elcograf S.p.A.

Hodder & Stoughton policy is to use papers that are natural,
renewable and recyclable products and made from wood grown in sustainable
forests. The logging and manufacturing processes are expected to co
nform to the environmental regulations of the country of origin.

Hodder & Stoughton Ltd
Carmelite House
50 Victoria Embankment
London EC4Y 0DZ

www.hodder.co.uk

In memory of Jessica O'Keeffe (1981-2021)

And for Rory and Ottilie

We move forward with hope and courage

The love I shared with her and the companionship we shared was the realest thing I ever did.
Former police officer Mark Kennedy, on a relationship he had while living undercover as activist 'Mark Stone'

If he was fictional, what did that make me, now?
'Alison', the long-term partner of 'Mark Jenner', a.k.a. Undercover Officer Mark Cassidy

Author's Note

Skylark is inspired by historical events. The Special Demonstration Squad's *Tradecraft Manual* (1995) is a real document, and was published by the Undercover Policing Inquiry in March 2018. It can be read in full (albeit with redactions) on the Inquiry's website: www.ucpi.org.uk.

The characters and organisations in this book are, however, entirely fictional.

DI Wells: Name?

UCO122: Daniel Greene.

DI Wells : Spell it.

UCO122: With an *e*. Greene with an *e*.

DI Wells : Where were you born, Daniel Greene?

UCO122: Barnsley, South Yorkshire, 14 October 1969.

DI Wells: Parents?

UCO122: Dorothy and Sidney Greene.

DI Wells: Dorothy's maiden name?

UCO122: Ellis. Dorothy Ellis.

DI Wells: Describe your parents.

UCO122: My mother Dorothy died when I was young. I was brought up by a stepmum, Diane, and by my dad. He worked down the pit at Woolley Colliery until it shut down. After he lost his job he hit the skids.

DI Wells: And how has that experience left its mark?

UCO122: I hate the Tory government for what it did to my dad and my community. And I've generally reacted against authority. Went off the rails as a teenager, ran away from home and went out to Spain working for some dodgy types. Ended up in prison there, briefly. Came out a couple of years ago, politicised. I've decided to put my energy into fighting the system.

DI Wells: Okay. That all hangs together. And your current employment?

UCO122: Aerial access.

DI Wells: Which is?

UCO122: I work on ropes, in building sites. Freelance. Which means I travel all over the country, change locations regularly, and spend periods of each week away from home.

DI Wells: You're something of a rolling stone.

UCO122: Very much so, yes. A restless character.

DI Wells: You'll fit right in. And your address?

UCO122: My current address? Er, Manor Road, Hackney.

DI Wells: Number and postcode?

[PAUSE]

No hesitations, Daniel.

UCO122: 142. Postcode. You got me. Okay – N16 4DX.

DI Wells: Right. Well done. Just make sure you have all those details properly memorised, in case anyone asks. You've already got all the tax records, birth certificate . . . There's just one final thing.

UCO122: What's that?

DI Wells: Your warrant card.

UCO122: Of course, I won't be needing that. Here you are, sir.

DI Wells: Thank you. And less of the sir. You're not a bobby on the beat now. Call me Martin. That's how we do it in this unit. We don't stand on formality.

UCO122: Right – thank you, Martin.

DI Wells: And remember: 'by any means necessary'.

I.

1996

3. PREPARATION

3.1 NAME

3.1.1. By tradition, the aspiring SDS officer's first major task on joining the back office was to spend hours and hours at St Catherine's House leafing through death registers in search of a name he could call his own. On finding a suitable ex-person, usually a deceased child or young person with a fairly anonymous name, the circumstances of his (or her) untimely demise was investigated. If the death was natural or otherwise unspectacular, and therefore unlikely to be findable in newspapers or other public records, the SDS officer would apply for a copy of the dead person's birth certificate. Further research would follow to establish the respiratory status of the dead person's family, if any, and, if they were still breathing, where they were living. If all was suitably obscure and there was little chance of the SDS officer or, more importantly, one of the wearies running into the dead person's parents/siblings, etc., the SDS officer would assume squatters' rights over the unfortunate's identity for the next four years.

Special Demonstration Squad,
Tradecraft Manual

1.

'So I think we're agreed,' said Rev, 'that it's high time we put on the biggest illegal street party this city has ever seen.'

Heads nodded around the circle. Skylark had counted ten. It was never many more, in those days, before the thing took off. Rev presided, perched on the stack of blue pallets they called the Throne, skinny, pale and hairless in a black boiler suit. Then the other usual suspects: Bendy Aoife, who was practising her splits on the floor; Mouse, his ankle-length dreads wound into a giant tam; Big Moll, in her flouncy floral dress and bovver boots. They were the core crew, the veterans, who had lived so entangled with one another in the tree-houses, benders and squats of the anti-roads campaigns that they could recognise each other by armpit odour alone. (Yes, they'd tried it.)

But as this Tuesday night was an open meeting, advertised in *Loot* and on a hand-scrawled notice pinned up in the anarchist bookshop, a few randoms had turned up, too: two old *Socialist Worker*-types with beards and flat caps, who had introduced themselves as Ken and Len; a keen-eyed student with braces and rainbow-striped trousers; a moody-looking guy in combats.

Rev wasn't keen on the open-meeting idea, would have preferred to keep the world-changing group small and closed, but Skylark had insisted: *This is not just about people like us! The better world has to be for everyone – it's that, or it's nothing.*

'We're ready to move up a gear,' Rev went on. 'Having tested our tactics and techniques twice now, first in Camden Lock, then on City Road, we know how to take a street and hold it. We know, more importantly, that our parties perform their function, that they give people a glimpse of a looking-glass world, a free world . . .'

Ken – or was it Len? – raised his hand, exposing the worn leather patch on the elbow of his donkey jacket. 'Excuse me, and apologies if I've misunderstood' – he frowned beardily – 'but I thought this was a meeting about a political protest. About bringing down capitalism.'

'Oh, absolutely,' said Rev, breezily. 'That's what we are all about.'

'So,' Ken/Len looked perturbed, concerned, 'why are we talking about parties?'

'A party is a portal,' Mouse responded solemnly. 'A flaming arrow of hope and life, fired into the heart of our dying city.'

Mouse was the world-changing group's spiritual guru, and when he spoke, which was rarely, he struck each word powerfully, like a gong.

Silence briefly fell. Ken and Len bowed their heads together and scratched their beards.

'To be clear: we are not about old, dull, dusty politics,' said Rev. 'We don't do placards, slogan-shouting, mani-festos, minutes, clauses or sub-clauses. I mean good luck with all that, if it's your bag. We just happen to think that changing the world should be a laugh.'

'A serious laugh,' added Big Moll. In the camps Moll had run kitchens, magicking heaps of mouldering swede into warming nutritious soups and curries, through mud, hail and snow. Once, during a long cold winter on Dolcup Hill, all the food in the kitchen had frozen solid, and she'd spent days hacking up vegetables with a machete. There was no messing around with Moll. 'No point lecturing people about how to change things,' she said, blunt as a spoon. 'We just get on and do it, and they can see for themselves.'

The rainbow-trousered student put up his hand. He looked very young, and his foot was jiggling nervously. 'I'm Jez. I, er, went to the party on City Road,' he said.

'It blew me away. It was magic. Really. I'm studying sociology, and I'd really like to know a little bit more about the theoretical underpinning . . .'

'Theo-what?' said Moll. 'It's a party, that's all.'

'Oh, okay. Thank you.' Jez seemed a little crestfallen, although his eyes still shone with eager idealism. But Ken and Len were not convinced. Len – or possibly Ken – now raised his hand. 'So the aim of your street parties is to stop the traffic?'

'The aim of our street parties,' said Bendy Aoife, in her soft Irish lilt, 'is to change the world.' As she spoke, her body arced over her left leg in an elegant, muscular curve. Aoife was training for the circus, which was as close to a career plan as any of them had got. Gazing for a moment at her goddess-like form, at the shining cascade of her red hair, both Ken and Len seemed to forget any worries they might have had, about anything, really. 'Cars are just a symbol,' Aoife singsonged, arcing now to the right, 'of everything that is wrong. They pollute, they divide people. They dominate our public space.'

Rev nodded languidly. 'We want to take things that have been sucked into the capitalist system, repurpose them, and give them back to people. We do it with streets, I do it with junk.' He waved his long thin hands at the room around them, which was his studio, under the arches in King's Cross. His sculptures, his machine beasts, peeped out from the walls and the nooks and crannies: small robots made from old oil cans; a spiny pterodactyl moulded from the insides of a grand piano. This was years before he got famous, but already Rev was making art from the things people threw away.

'It's a plain fact,' he went on, 'that every road in this country has been commandeered by the car and oil industries. And people don't even realise! They don't question the fact that they can't walk, or meet, or let their children play in a public place. They think this is normal . . . Until, that is, we put on a party, and show them what a street could be.'

Ken and Len hummed and nodded and seemed, at least temporarily, satisfied.

'So let's see.' Rev jumped up with the air of a man who momentarily wished he were mainstream enough to own a flip-chart. 'Our first action, Camden Lock – Moll, how many punters did we get there?'

'About five hundred, give or take.'

'And in City Road?'

'The cops said a thousand. I'd say three,' said Moll.

'So,' Rev went on, pacing to and fro, 'with a growth trajectory like that, if we play our cards right, next time we could be looking at a crowd of – what? – ten thousand?'

Aoife inhaled a whistle. 'Okay,' she said, 'we're going to need a big road.'

'The biggest!' cried Rev. 'Which is where I hand over to my warrior queen, Skylark McCoy.'

Skylark stood up, smoothing her ragged ballgown over her intensely patterned leggings. Her energy was as springy and sprightly as the matted blonde hair that tumbled around her head. In her hand she held a battered black notebook, instantly recognisable to the veterans present. This same notebook, now held together with gaffer tape, had accompanied them all through years of anti-road actions, from the encampments of Oldbury to the tree-dwellings of Dolcup Hill and then to Harfield Road, the squatted east London street where they had resisted the construction of the M11.

She looked young, had something girlish and innocent about her, but aged nearly twenty-one she had done her time, and earned her stripes. Her face was open, transparent almost, and her world-changing passion shone through it, pure and clear.

She flicked to a page at the back of the notebook. 'The plan for this summer is,' she said, 'to take over the motorway. The M41, in Shepherd's Bush.' She produced from the notebook an enlarged photocopied page of the London A–Z, a thick section of road highlighted in yellow. 'We've chosen this location for a number of reasons: less than five minutes' walk from tube and train stations, so easy to get large numbers of people into and out of. As it's a bridge, here' – she pointed at the map – 'there are only two points of entry and exit, which makes it easier for us to take and hold the space. Also, it's highly disruptive. It leads to the Westway on one side and the Holland Park roundabout on the other. It was originally supposed to be part of a ring of major roads around the whole city, until – thanks to certain vigorous campaigns – the government quietly shelved its plans.'

'The Mecca of the car,' remarked Mouse, with a faraway smile.

'Quite,' said Skylark. 'So . . . what do we think? What should this party look like? What do we need? Any ideas about strategy? Let's throw everything into the pot now, get this thing cooking.'

'A big fat sound-system,' said Rev.

'Obviously.' Skylark nodded. 'I've been talking to Carl about that.'

'A sandpit,' suggested Moll. 'For the kids. A giant sandpit. Right across the central reservation.'

'Yup,' said Skylark, jotting now in the black book. 'Let's look into sand.'

'For decor I was thinking of giant fluorescent alien heads,' said Rev. 'And perhaps some giant fabric sunflowers.'

'We'll need banners,' said Ken/Len.

'Making the anti-capitalist nature of this event clear,' added Len/Ken.

'Of course,' said Skylark. 'That can be your area, Ken and Len. Anti-capitalist banner-making.'

Aoife, who had been in a headstand throughout this discussion, brought her feet down elegantly. 'Maybe a dance troupe could depict the downfall of capitalism through expressive movement.'

Skylark scribbled in the book. 'A dance troupe. Noted.'

There was a brief and thoughtful silence before Rev spoke again. 'Our biggest problem,' he said, 'is transport. Obviously the last couple of times we've relied on Dave and Ali and their lorry. But they've gone off to Ireland now, return date unclear. We could hire vehicles, obviously, but it's all money . . .'

He was interrupted by a soft cough. The moody guy in combats, who had until now been smoking a rollie and watching the proceedings in silence, was holding up his hand. He cleared his throat again. It struck Skylark that maybe he wasn't so much moody as shy. 'I might be able to help,' he said.

He had a northern accent, buttery rounded vowels. Something else struck her about him, too, a kind of familiarity. His eyes were dark, and so was his hair. He had on a neat checked shirt, short-sleeved, and she noticed in particular his arms, which were weightlifter strong. You didn't often get arms like that on the world-changing scene: most males tended towards a scrawny vegetarian skinniness. She would have liked to spend a while looking at those arms, just taking them in.

Rev raised a hairless brow. 'Oh?'

'I'm Dan, by the way,' he said, took another tug and breathed out smoke long and slow. His eyes met hers and she looked quickly – too quickly, dammit – back into the depths of her notebook.

'And what brings you here, Dan?'

He shrugged. 'I saw your ad in *Loot*, and it – intrigued me. Thought it was about time I changed the world.'

Skylark smiled at this; the others didn't. Rev, despite the movement's firm commitment to laughs, did not actually like to joke about this stuff. She could sense he was about to be arsey, as he sometimes was with newcomers, and she wanted to pre-empt it, not only because of the arms but also because Dan seemed not quite their usual type, not a hippie, more ordinary. They needed to make people like him feel welcome.

'Nice to have you here, Dan,' she said quickly. 'You were saying you could help?'

'Yeah. I'm pretty handy,' he said. 'I can lift stuff, build stuff. And if you're in need of transport, I have a van.'

With these words the atmosphere in the room relaxed.

There was no better route to Rev's affections than owner-
ship of a functional van. He hated having to rely on Dave
and Ali, Traveller mates of Moll's, who had a habit of
lunching out at crucial moments, leaving people and tat
stranded all over London. A man with a van was most
certainly not to be sneezed at.

'Well, Dan,' said Rev, his pale lashless eyes now sending
out noticeable gleams of welcome and enthusiasm, 'the
pleasure is ours.'

Later, she wouldn't remember exactly what they'd
discussed that day. As well as the logistics for the M41
action, the one destined to put their world-changing
group on the map, they might have touched on the
grander philosophical picture. Topics such as: whether
capitalism was compatible with fundamental human
dignity (hell, no), whether mainstream feminism was
compatible with true feminism (debatable), whether their
movement was replicating the macho patriarchal norms
of capitalist society (yes, according to the three women
present, and acknowledged with varying degrees of reluctance by the male members of the group), whether there
was some kind of inner vibration within rave music that
led its listeners inevitably towards freedom, love and
harmony (yes), and whether they should make a final

decision about the location of the action planned for that summer by means of consulting their shamanic spirit animals (proposed by Mouse, and left undecided).

The thing she would remember was that at the end Dan waited for her by the door. 'Hey,' he said. 'Thanks for helping me out back there. I'm not much of a talker.'

Her face was all sunshine and birdsong. 'That's okay,' she said. 'It can be scary, I know, when you're new.'

'It's not just me, then.' He looked down, with an odd slanty smile, at her fractal-print leggings, the crushed velvet ballgown, the unlaced boots. His smile revealed gappy front teeth, which she noted and approved. She shifted her weight, suddenly self-conscious, which she wasn't usually. 'I'm sure I'll get used to it,' he added.

'So you're coming back?'

'If you'll have me.'

'Of course.' This came out a bit too quickly, and maddeningly, she felt red rising up her cheeks. She stepped hurriedly past him, out into the street. Turned to wave as she walked away. 'So, see you next week. Right?'

He smiled, raised his hand, and she wondered whether he was shy, or slyly arrogant, or both. 'Right.'

2.

I T was ironic, she thought, as the lift clanked up past floors one and two, to feel more can-do about changing the world than about changing your own life. By floor three, the buzzy positive energy of the meeting had slipped away, and something grey and stifling settled over her, like someone else's too-big coat.

She pulled back her shoulders and stood up straighter, trying to shake it off. Fourth floor. Today might be a good day. Maybe Mikey would have been out job-hunting. Maybe he would have made some progress with the decorating. Maybe, when she opened the door, he would greet her with a cheery smile.

It wasn't impossible.

And – five. The lift shuddered to a halt and the doors wheezed open. Everything in Heron Court was a bit knackered, a bit battered, from the graffitied stairwells to the black bins piled high with bags, from the play-ground with its rusting slide to the tiny caged-in football pitch, which rattled loudly with every stray ball. She had only moved in six months ago and was still getting used to this settled London life. She knew how lucky she was to get a council flat, and in Hackney, too. But she couldn't help missing the squats and camps, the company, the space and fresh air. Even with the windows wide open, there was never enough air in the flat for her.

She lugged her blue bags of shopping down the walkway, fished her key out of her pocket and shouldered open the front door. The tins of paint were still sitting unopened in the hall. The peeling flock wallpaper was untouched, the grubby lino intact.

So much for progress with the decorating.

Mikey was in the living room, stretched out on the beanbag he had found in a skip and presented to her as a moving-in gift. His feet rested, along with three empty Stella cans and a brimful ashtray, on the cardboard box they had been using as a coffee-table. Around him a light scattering of papers, bits of baccy, roaches. The whole room was stuffy with weed; just breathing got her light-headed. She dumped the shopping bags in the kitchen, then crossed into the living room to throw open the door to the balcony. A blast of cold February air cut through the fug.

'All right, darlin'?' he asked blurrily.

She bent down to kiss him, tasting stale smoke. 'Yeah.

Not bad.' She ruffled his long matted hair; he squeezed her hand. His eyes had the blank, black-hole look. She could see he hadn't done anything but drink, smoke and watch telly since she'd left for work that morning. Frustration ballooned in her chest, but she took a deep breath and compressed it, made it small and unobtrusive. One thing she had learned was that it really wasn't worth arguing when he was blitzed. She left him to it, went to sort out the dinner.

Hahaha hahaha.

Through the open kitchen door came the tinned laughter from the TV show Mikey was watching. It was that American sitcom, the one about the blonde woman with the square jaw.

Oh, Cybill, what shall I do? I'm so depressed, drawled a voice from the telly.

Do what I do, another replied. *Deal with it head on, years later.*

Hahaha. Hahaha.

'Want some food?' she called, getting tinned tomatoes, garlic and sardines out of the bags. She had heard that oily fish was good for a low mood.

'If you're making it.' Mikey used to love cooking: in the squats and camps he'd been Moll's sidekick in the kitchen, sneaking Scotch bonnets into the curry when she turned her back and blowing everyone's socks off.

But cooking was something else that had fallen by the wayside, since he'd gone downhill.

She had asked herself, many times: should she have seen it all coming, the weed, the booze, the blank-eye thing? Done something more to help? Been firmer, been kinder? 'The thing about Mikey,' Rev had said once, back in the days when they could still talk about him without arguing, 'is that he doesn't have back-up.'

And that was it. Plenty of people came to the world-changing life because they wanted an adventure, or didn't like the other options. Mikey had never had other options. He had no family in the country, since his mum had retired from social work, sold her place in Birmingham and moved onto a boat somewhere in deepest France. ('I think she wanted to get away from people,' Mikey said. 'Including me.') When they had been evicted from their last squat, in Harfield Road, she and Rev and most of the squatters had been pushed straight to the top of the housing waiting lists, partly because the anti-M11 campaign had been such a shitshow that the powers-that-be were desperate to shut them up.

Mikey had been offered a flat in Brent, but for reasons that were equal parts principled and self-defeating, he had turned it down. His life, he claimed, would be lived on the road. He had been planning to head over to Ireland with a convoy, but the old Dodge 50 he'd had for years had finally packed up, and he had no cash for another vehicle. 'Just let me stay for a bit, Sky,' he'd said. 'Until I get myself sorted. I'll help you do the place up. Take care of you. You don't want to be rattling around by yourself.'

Take care of me, she thought sadly. Right.

'So, anyway, it was a good meeting tonight,' she called, as she sliced garlic, heated oil. 'Everyone's up for this motorway action in the summer. Rev thinks it's gonna be big.'

'Oh, yeah?'

'Do you think you might help out? Everyone always asks after you.'

That was a lie. People had stopped asking after Mikey, not because they didn't care but because there was never anything new to say. Still drinking, still stoned, still a burn-out. She poured the tomatoes into the pan, closed her eyes and summoned a mental image of how he used to be.

Mikey had built the best tree-house in Dolcup Woods, three rooms, with a skylight over his bed to watch the stars. They had slept there through the bitter winter, clinging to each other for warmth. He had an unmatched knowledge of the natural world, made her strange medicinal concoctions with twigs and herbs when she got a cold, taught her to identify a tree through the taste of its leaves. He showed her how to wash using one cupful of water, and to breathe air directly from his mouth while he moved inside her at night.

Mikey, that old, beautiful Mikey, had given her – what did he give her? – a taste for freedom.

She put the dinner on two plates and carried them into the sitting room. Put them down on the cardboard box, sweeping away the cans and roaches to make space. As she sat down, something shifted in her mind: she felt she was watching the scene from above, seeing it from a different, outside perspective.

If you'll have me.

Of course.

She saw the bare bulb hanging from the ceiling, the crappy beanbag, leaking its polystyrene beans all over the lino. The cans, the rubbish, the smoke. The boyfriend so stoned he could hardly talk. She had never wanted a two-up-two-down kind of life, but there was nothing poetic about this. It felt – the word came to her with shocking clarity – squalid.

She put down her fork, took a deep breath. 'You didn't do any painting, then,' she said.

'Don't start. Not now,' said Mikey, with his mouth full. He turned the volume up on *Cheers*.

'Okay not now, but when?'

There was a crash that sent her jumping out of her

skin. He had thrown the remote at the wall. A few centimetres to the left and it would have caught her head.

'I told you, Sky, just leave it out.'

So she did: she left it out. She stood up and walked down the hall to the bedroom because there was no point. Sat heavily on the bed. There was an envelope lying on the pillow, and she opened it, for a bit of distraction. It was a BT bill. She scanned it absently, then caught her breath.

'Mikey?' she called.

'Yeah?'

'How come we've got a three-hundred-pound phone bill?'

After too long, he said, 'No idea.'

She turned the sheet over and looked at the columns of itemised calls. There was an 0800 number that came up repeatedly, almost every day. The phone hung on the bedroom wall. She picked up the receiver and dialled it. Two rings, and then a breathy woman's voice answered.

You have dialled HotLine. The hottest chat, with the hottest girls.

Slowly, she put the receiver back into its cradle. The rage she felt was pure, a shot of adrenalin. She walked into the living room and held the phone bill in front of Mikey's face. He batted her away.

'You've been calling a sex line.'

Was she imagining it, or was there a hint of a smirk on his mouth? Her insides were cold steel.

'You've been calling a sex line, on my phone, in my flat, while I've been out working to pay your rent.'

He shrugged. 'Relax, woman. Just say the word, and we can call it together some time.'

His blank, heavy eyes met hers. Slowly, she screwed up the paper in her hand. 'You can't do this. I'm not having it, Mikey. Not with you like this. I need you to leave,' she said.

'*You need, you need*,' he said, mocking, and then time folded in on itself as he jumped up off the sofa and grabbed her by the shoulders. He pushed her back across the room and out of the door onto the balcony. He had her up against the rail, and his hands had slipped upwards, from her shoulders to her neck. Mikey was skinny, but he was strong, all muscle. He leaned into her, squeezing out the air.

After he let her go, she stood there for a moment rubbing her neck, catching her breath.

'What is happening?' she breathed. 'You could have . . .'

He kissed his teeth. 'Fuck's sake,' he said. 'You didn't think I was actually going to do it, did you?'

3.

M IKEY moved out that same night. She heard him shuffling around as she lay in bed not-sleeping, and by the time she got up (late, jaded) for work, he was gone. He had even tidied the living room and left a tenner by the kettle, with a note underneath. *For the phone. Sorry. Mx*

Standing there that cold winter morning in her nightie, amid the stark empty stillness of the flat, she pressed her fingers into the corners of her eyes to stop them leaking. Mikey was gone; she was on her own. It had been a long time. In the squats and camps you were never alone: there was always someone busting in on you in the bog, swearing in the background while you phoned home, rifling through your coat for filters or your last square of Dairy Milk.

She hadn't been on her own since Henfield, when aloneness had been a fundamental condition.

Padding down the hall, she looked into the mirror.

There was a sore patch on her neck where he had grabbed her, slightly raised and greenish, but probably too faint for anybody to notice. She still dressed carefully, covering it with a scarf. It was just the kind of thing you didn't want to have to explain, not at work, in a new job.

That day she focused on thinking about arms. Specifically, big strong arms, which could enfold her completely. Arms to keep the world at bay. She thought about them in the bus, giving herself an imaginary hug as it chugged past King's Cross and up Caledonian Road towards the play scheme. She thought about them while she was eating her tuna sandwich at breaktime.

'Earth to Sky,' said Suze, catching her staring at the wall. 'You all right, chicken?'

'Probably,' she said, before she could think of a better answer. She was sitting at the desk in the tiny office, the only place in the play scheme where anyone could be guaranteed ten uninterrupted minutes. It was piled with Suze's admin, half filled-in referral forms and insurance reminders, unopened policy documents and case files stained with bits of lunch. 'Just – oh, life, you know.'

Suze's dark eyes glimmered behind her glasses. When it came to people, Suze didn't miss much, which was a good and a bad quality in a boss. 'Need to go home?' she asked.

'Nah.' Skylark waved away the suggestion. 'It's all good. I'll be fine.'

'If you're sure.' She nodded. 'Now, where the fuck's that letter?'

There was an overpowering smell of some exotic perfume – sandalwood? – as Suze leaned over her to rummage through the piles of paper on the desk. Admin was not, Suze would cheerfully concede, her managerial strong point. It accumulated in piles, which turned into drifts, which were eventually bagged up by some despairing cleaner and put out with the recycling. Somehow, it never seemed to matter. None of the normal rules applied to Suze.

'Ah!' she exclaimed, extracting a curry-smeared piece of paper. She pushed her glasses up her nose and adjusted her flowing skirt, which was hot pink with an orange pom-pom fringe. Suze's clothes were always searingly bright; people joked that she never got pulled up on paperwork because the head of Children's Services couldn't look at her for long enough.

'Here it is. There's some politician coming to look around. An MP.'

'Really?'

'Yeah. Someone from Labour, the shadow minister for families. They're promising Children's Services lots of money if they win the election next year. And this is a photo opportunity, yadda yadda.'

'Do you think they will?'

'Of course,' said Suze. 'Absolutely.'

Suze was big on the power of positive thinking. *The kids here have many disadvantages, but they can have a good life*, she had explained, on Skylark's first day. *Our most important job is to believe that.*

Skylark had got the job at the play scheme a few months ago, after moving into the flat. When she'd lived in the camps she'd just signed on. Everyone did. But after moving into the flat it seemed like time to grow up. She'd heard about The Crew, a scrappy little charity off Caledonian Road, from Bendy Aoife, who was an old friend of Suze.

She had no experience and there hadn't been an interview, as such. Suze had shown her breezily around the centre, with its tiny jumbly office, its soft-play room, and its main area scattered with books and art supplies. The whole place smelt of baked beans and piss, which she found oddly reassuring. She'd turned her back on institutions early in life; it was quite nice to be back in one.

The whole place was mayhem, though: overcrowded, small and sweaty and loud. Suze sailed through the sea of children, like a galleon with hot-pink sails.

'Have you ever worked with kids before?' Suze had asked that day, fixing her with those eyes.

'Never,' she had answered, because honesty was the best policy.

'Got a good intuition?'

'Yeah,' she said. 'I think so.'

'Well, use it,' said Suze, 'and you'll be fine.'

She only got ten minutes for her sandwich, before getting back to Jase. She couldn't leave him for long because he tended to eat stuff – his preference was for brightly coloured plastic: paint pots, toy cars. Once, his mum had taken him to hospital when he seemed to have bad indigestion: they'd pumped his stomach and found the remains of all her credit cards.

'Wanna do some drawing?' she asked, but Jase looked through her. He was a slight eleven-year-old with yellow-brown skin and long black hair that his mum plaited down his back, like an Apache chief. They called him Shaman because he had a faraway look in his eye, a wide pure gaze that seemed fixed on some invisible horizon. A long string of saliva hung from one side of his mouth. 'No? What do you want to do, then?'

Jase reached out and took her hand firmly. She knew full well what he wanted to do: the same thing he always did. Crossing his legs, he rocked backwards and forwards. She gave in, as usual, and rocked backwards and forwards too, in the same pendulous motion. He rocked harder; she did too. He made a loud clicking sound with his mouth; she copied him. Jase laughed joyfully and did it again.

She fixed her eyes on Jase's, and carried on copying his every move. He was definitely on to something, she thought, as she rocked. The back-and-forth motion was really soothing. There was something relaxing about sitting together, moving together, without any words. She'd had enough of words, sometimes.

You need, you need.

Mikey's voice echoed in her head. Had he always had that meanness in him? Or was mean Mikey the veneer, his real, beautiful self having been clouded and obscured by booze and weed? Which way around was it? The better you knew someone, she thought, the more mysterious they became, the more layers they acquired, the more difficult it was to say who they really were.

'Bah,' said Jase, pausing mid-rock and giving her a glorious unencumbered smile. 'Dah.'

She realised, with a rush of what she could only call love for him, that Jase was his own true person. She took his hands. 'You should teach us all,' she whispered, 'teach us not to be so fucking complicated.' But he just stared at her with his faraway gaze.

'And this here is Sky,' Suze's voice boomed, from

overhead. 'And Jase, who has been with us since he was two. You're one of our lifers, hey, Jase?'

Jase turned the gaze up towards Suze, who was standing behind them. All the kids loved Suze – she was like a superhero to them, all cartoonish colours and charisma. Standing next to her now was a woman with an expensive haircut, dressed in a black trouser suit with a silk scarf around her neck. She had a large red rosette pinned to her lapel.

'This is Tessa. Say hello, now.'

'Hello, Jase.' Tessa nodded at both of them, held out her hand. 'So nice to meet you.'

Jase said nothing, left her hand hanging there in space. He was gazing at her lapel.

'Tessa wanted to get a photo – do you want to sit down next to them, just there?'

There was some to-ing and fro-ing with the camera and Tessa crouched next to Jase on the floor. They all smiled for the camera, except Jase who, just as the photographer took the shot, leaned over and sank his teeth hungrily into the shiny red rosette.

4.

THE next Tuesday, when Dan walked into the meeting and sat down beside her, some deeply buried muscle between her shoulders relaxed. She liked his broad body, his gappy front teeth when he smiled hello. She liked the way he looked, clean and fresh and together. He was a still point in the room.

'You're back,' she said, trying not to sound too relieved. Around them the other members of the world-changing group were nattering and smoking and foraging for biscuits. Ken and Len were trying to work the kettle; Jez had cornered Rev, and was talking to him earnestly, using words including *dialectic* and *liminal*.

'You sound surprised,' he replied.

'I thought we might have scared you off.'

'No. Although I was a bit worried about seeing those again.' He nodded at the fractal-print leggings.

She stuck out her tongue, mock-offended. 'I'll wear navy blue next time. Just for you.'

'I'd appreciate that.'

Rev, who had shaken off Jez and assumed his customary position on the Throne, called the meeting to order. There were big decisions to make, he announced.

'We need to divide into crews, assigned to each vehicle,' he said. 'So far, we have two trucks. I'm driving one with Carl's sound-system and the sand for the sandpit, plus miscellaneous decor. The other is Dan's van, with the Spirals' sound-system and the tripods. Then there is the car we need to create the crash.'

'The crash?' said Len, looking confused.

'That's how we block the northbound side. We did the same in Camden. One of us – someone brave – drives an old banger slowly up the M41 to this point here,' Rev pointed at the map, 'and crashes across the lanes deliberately. Bingo! The traffic grinds to a halt. Meanwhile, the rest of us leap into action, unload the tripods, whack them up quicktime, and get people sitting in hammocks at the top. Once the sitters are in place, our tripods are extremely difficult for the police to remove.'

'But what about the southbound side?' asked Moll.

'We let the crowd do that. It's people power. With the numbers we're expecting they'll completely block that side as they leave the station.'

Bendy Aoife was talking while practising her pigeon

pose and simultaneously constructing a roll-up. 'So I had an idea,' she said. 'Perhaps after we take the road and get the sound-system going we could have a Marie Antoinette character roaming through the crowd. You know, tall grey wig and huge skirts. On stilts. Let the peasants eat cake, and all that.'

Rev nodded keenly. 'I love it, Eefy,' he said. 'Could we have people hiding beneath her skirts, perhaps?'

'Drilling,' said Mouse, who was sitting cross-legged on his chair, the ends of his dreads trailing on the floor. He didn't open his eyes as he spoke. 'Drilling up the road. Planting trees.'

'That's it!' cackled Rev. 'The forces of destruction and creation, hidden from the fuzz beneath a huge ballgown!'

The plans for the motorway party were swirling and forming a whirlwind. This was the magical feeling, the fizzing kinetic energy, when a collective world-changing idea began to take shape. It was as though the thing assumed a life of its own: a glorious event already existed in concrete form somewhere in the future. All it needed was for each of them to accurately perceive their part in it and play it truly and honestly.

Either that, or it would be a complete shitshow and they'd all get arrested.

'Right,' said Rev, clapping his hands. 'Everyone on your

feet. We need to practise until we can assemble this tripod in less than five minutes.'

'It's complete chaos, but also weirdly efficient,' Dan said, as they walked together down York Way towards the station. It was cold, tiny flakes of ice twitching through the air.

Skylark put up her hood and hunched into her bright orange puffa. She laughed. 'That sums it up.'

'Do you think it will all actually happen?'

'I have no idea. If the universe wants it to.'

Now he was the one to laugh. 'Do you believe that?'

'I know it,' she said. 'That's what it was like in the camps, too, or in the squats, when they came to evict us, and it was battle stations. Nobody has a clue what's going on, but somehow you all get into a flow. You need something, and before you even ask, somebody has given it to you.'

'Spooky.'

'No, I don't mean hocus-pocus.' She dismissed him with a wave of her hand. 'It's just real. It's what happens when a group of people have a common purpose. When they're of one mind.' She looked at him, grinned. 'You'll learn.'

'Are you patronising me?'

'You will,' she said. 'You've got stuck right in.'

During the meeting Dan had volunteered, to Rev's

obvious delight, to drive his van from Point A (probably a squatted warehouse in Dagenham) to Point B (an industrial estate near the M41, where he would wait for the road crew) and then Point C (the party area, the final destination). He had offered to support the banner-making subcommittee, and take over the production of flyers from Ken and Len, for whom double-sided photocopying had presented a technical challenge too far. He had also said he would put the names of people who had been to their meetings into something called a database. (It was just like a list, he explained, when this met with blank looks.)

'I thought I might as well throw myself in,' he told her now.

'This is your first direct action?'

'In a way. My first for a long time.'

'And why now?'

He frowned down at his feet, then looked up with his slanty, gappy smile. 'I'm changing my life,' he said.

'Nice one,' she replied. 'Me too.'

That evening a gloomy winter chill lay over the whole zone, the long stretch of road illuminated by passing headlights. This was still old London, before the dawning of the sourdough age, when the streets of King's Cross were filled with scowling men of questionable motives and skinny, dirty-faced women waiting brazenly on corners in too-short miniskirts. She noticed how safe she felt in the presence of this man. It wasn't just his size, but something about the way he carried himself, like he owned the space.

They had reached the station, and stood for a moment as cars rushed past them along Euston Road. Dark figures criss-crossed between the tube and the train, busy, going places, hunched into coats and hats. A square of greenish yellow sloped from the newsstand.

A 38 bus had just pulled up, and passengers streamed out of the open back door.

'This is me,' she said, by way of goodbye. But then, seized by some breathless impulse, added, 'Unless you fancy a pint.'

He looked a little surprised, but said, 'Yeah. Why not?'

They went over the road to the Auld Triangle, a sticky-carpeted boozer of the kind that existed back in those pre-smoking-ban days, when blackened shrunken lungs were considered a perfectly acceptable feature of life's rich tapestry. It was a ramshackle old place, just a slot machine blinking manically and a cluster of old men nursing pints at the bar. An Irish flag hung on the wall, next to a flickery old telly showing *Emmerdale* on silent.

She sat down while Dan got the drinks: a cider-blackcurrant for her, a Guinness for him.

'So sorry, what's your name?' he asked, and she told him. 'No, but I mean your real name,' he said.

'That's it.'

'But nobody's called Skylark.'

She shrugged. 'You can call me Sky. Everyone does.'

'Sky.' He tested this out, seeming to find it more acceptable, but still shook his head and chuckled into his pint. 'Sorry. I'm still getting used to all this.'

'A lot of us chose our names,' she explained, 'when we were in the camps. Partly it's a security thing. It makes it harder for the police, when they arrest you. And partly we wanted names that suited us.'

'Like Rev?'

'Yeah. I mean you can see why, right?'

Dan nodded. Rev had lost his hair when he was a teenager, after a car accident. It was his pale, skinny baldness, his lack of eyelashes and eyebrows, his black clothing, his unusual but authoritative presence: he looked like a priest – albeit one from another planet.

'And you?' he asked. 'Why Skylark?'

'People say I have a clear voice.' She had taken a large swallow of cider and black, its warmth spreading across her insides. 'So,' she said, leaning forward, 'you first.' He frowned. 'The story, of why you're here. The life change.'

'Oh, that. Really? You want to know?'

'I always want to know what brings people in.'

He had his smoking stuff out on the table and she noticed that his big hands shook a little as he assembled the necessaries. Was he nervous? Of her? There was a pause, quite a long one, as he filled the Rizla, rolled it, smoothed it. He stuck it in the corner of his mouth and looked her square in the eye. 'I got myself into some trouble, a while back,' he said. 'Did some things I shouldn't have done.'

'Such as?'

'Nothing special. Petty crime.' He lit his fag, exhaled. Seemed to be considering whether to say any more. 'I was really lost.' Smoke wreathed his head, tendrils reaching for the yellow-stained ceiling. 'I grew up in a very political environment, you see. My dad was a miner. An electrician, actually, but he worked down the pits. So, yeah, back in the eighties, when I was growing up, everyone was out on strike – everyone hated the police, the government. You couldn't avoid politics – it was like air, you just breathed it in.'

She was too young and too southern to remember it, had only vague memories of the miners' strike, of Mother tutting, *What is the country coming to?* as men shouted at each other on the telly.

'But – yeah. Life went on, and I lost that sense of the bigger picture. Got totally bogged down with all the other stuff, life stuff, my dad's drinking, my own problems. I just became a very angry person. I put myself in bad situations. It was only after things went tits up that I thought, who am I actually angry with? It's not a person.

It's the system. It's the system that did this, to Dad, to me, to all of us.'

'Rage against the machine.'

'Too right.'

There was a blast of noise from the flickery old telly. One of the old guys had raised the volume; the others had lifted their faces from their pints to watch. It was a relief to get a distraction from the intensity of the conversation, to look away from Dan towards the screen. A familiar face was speaking directly to camera. He was saying, with precise consonants and a tone of complete certainty: *The entire historical context of the IRA's position has been overtaken by history, and only they seem not to notice it. Sinn Fein must now play by the rules of democracy, or not at all.*

One of the old guys frisbeed a beer mat at the telly. There was gruff laughter from the others; behind the bar, the slow-moving barman shook his head over the taps. The grumbles of dissent continued until the man was gone, replaced on the screen by a clock graphic.

That was a broadcast by the Leader of the Opposition, the Right Honourable Tony Blair, MP.

'Perhaps we should leave,' said Skylark. 'Go somewhere we can hear each other.'

'Ah, I was pretty much done anyway.' Dan waved his hand again. 'That's my story – the abridged version.'

There was a blast of pompous news music. 'Now, boys, enough of all that.' The slow-moving barman had made his way over to the telly, where he flicked the channels until he found something more innocuous: a writhing man in silver trousers singing *Spaceman, I always wanted you to go into Spaceman* . . . The old guys lost interest, turning as one back to their pints.

'I'd like to hear more of it some time.'

'Yeah, maybe.' He shifted in his seat. 'And how about you, Skylark? How does a young girl like you find herself mixed up in all this?'

'Oh, that is a story.' Their eyes met and something passed between them, potential, or maybe just possibility. They sat there for a moment, taking each other in, unfamiliar beasts from faraway planets. Behind them, the slot machine rattled; the barman slowly topped up another Guinness.

'I bet it is,' he said. 'Tell me.'

5.

I T was the August bank holiday of 1992 when the Travellers moved onto Henfield Common. Local feelings were given voice by Sally Reynolds next door, in her interview for the TV news.

'They should bring in the army,' she said, as the reporter nodded sympathetically. 'It's devastation, absolute devastation. The noise, the drug pushing . . . and *the smell*. We'll be glad to see the back of These People, and everything they bring with them.'

Small, square, boxy Henfield, previously best known as south-east finalist of the Village in Bloom Award 1989, had found itself unwillingly at the centre of a national news story. Hundreds of dirty, colourful buses were parked on the common, and thousands of dirty, colourful people had set up camp. These People filled the village streets, and they were awful, terrible, with ragged clothes and unbrushed hair and untrained dogs. They were, according to the reporters who arrived hot on their heels, not only

Travellers but something even worse: New Age Travellers. 'Not so much picturesque horse-drawn caravans, but rusty trucks and rave music,' explained Mother's *Daily Record*.

Over four days, a large shanty town of tepees, stalls and marquees sprouted and sprawled, right there on the quiet, undulating field beyond the cricket pitch. It was abundantly clear that These People didn't care about cricket or, in fact, about anything normal. They stole booze from the Village Stores, their dogs had savaged six sheep, and they kept the entire village awake all night with the *dum chick-a dum, dum chick-a dum* of their music, which shivered maddeningly, endlessly out into the summer air. The local police force was so outnumbered its officers could only look on.

'We simply don't have the manpower to break up a gathering of this size. There are upwards of eight thousand people here,' Superintendent Blakeman had told the reporter, while next to him, Sally Reynolds's beaded earrings trembled with righteous indignation.

'It's a *scandal*,' breathed Mother, deliciously, when they watched this on the TV news. They were sitting on the settee at the time, eating dinner (they always ate dinner in front of the telly, then, so as not to notice the gaps at the table). Tick-tock, tick-tock, from the clock on the mantelpiece, Mother's brass carriage clock, which she had arranged with the two baby pictures, one on either side: the baby on the right was Janie, pinky and plumpy and only two days old, the one on the left was serious and frowning and weight-of-the-worldy.

The serious baby was Lilian – she liked to be called Lily, then: she wasn't Skylark yet – and she was sixteen-and-a-bit. She would never have articulated to Mother the feelings that the arrival of the Travellers had stirred up in her; she would not have been able to categorise the internal stirrings she felt as she walked past the common as excitement or fascination, as hope or – was it? – even longing.

It went without saying that Mother – Mother, with her *Daily Record* and her Tupperware collection and her strong views on parking – would not have understood. And it was important to Lily-aged-sixteen-and-a-bit not to upset Mother, not to disturb her delicate emotional balance or make anything harder than it already was.

But:

'Come on, what are you scared of?' On the way home from school, she started on at her friend Rupert, telling him they should go to the common, see what was going on, wander a while amid the tents and trucks, soaking up the extraordinary atmosphere.

'We can't go now! Not like this!' he cried, gesturing at their clothes. (Him, a knee-length velvet coat and flares; her, an oversized man's jacket, customised with a silver ribbon at the waistband and a 'Ban the Bomb' badge. They were the only teenagers in Henfield who didn't wear Burberry, the only ones who didn't hang around outside McDonald's in Haywards Heath on Saturday afternoons. They were, everyone else made clear, the weirdos.)

'Why not?'

'We don't look suitable!'

She laughed. 'You sound like Mother.'

Rupert reprimanded her with his pale blue eyes. Not only was Rupert unlike other people in Henfield, he was unlike anyone, anywhere. For a start, he was incalculably posh – he lived at Elsmore, the big manor house at the top of the village, up a massive driveway. But he was also skinny and pale and effeminate and, since his accident, had no hair at all. His white, gleaming bald head was like something from outer space. Everyone at school called him the Alien.

Everyone apart from Lily, whose whole house was about the size of Rupert's front room, but who loved him like the sibling she didn't have.

'Okay, then,' she said. 'Go home and change. We'll head over there later.'

There was some activity going on behind Rupert's eyes. She knew he was planning his outfit. 'What are you going to tell her?'

'Mother? I'll tell her I'm at yours.' (This was a safe bet, as Rupert's parents didn't generally answer their phone: it was in a dark and far-flung bit of the enormous house they called 'the pantry'; nobody ever heard it.) 'What about you?'

Rupert shrugged. 'You know Isabel and Robert. They don't care.' According to Rupert, his parents – he called them by their first names – were Bohemians. Without knowing quite what the word meant, Lily felt it suited them exactly. Kind of ornate and foreign, like the woven rugs and cushions and heavy brass candelabra that filled Elsmore's dusty, cavernous rooms, like Isabel's long French cigarettes, and Robert's battered Panama hats. Rupert's parents terrified and fascinated her: they might have lived in the same village but they were from another world, a world that, no matter how much she loved Rupert, she could never be part of.

As soon as they came up with it, the plan felt somehow inevitable. Of course there was a huge Traveller camp on their village common; of course These People had chosen Henfield. They had come – of course they had – for a very specific reason: because they knew that Rupert and Lily needed them.

'Meet you outside the Stores at eight?'

'I think I'll wear my biker boots.'

* * *

The whole vast rambling encampment smelt of wood-smoke, from the fires and braziers, and petrol, from the generators. And there was another smell too, something like fireworks, Lily thought. The smell of excitement, of danger. The heady summer air drew her deeper into the heart of it, past stalls selling earrings and tie-dyed clothing, past sound-system marquees where people nodded along to reggae and trance. Everywhere there were small groups of vans and tents, circled around fires, people sitting on upturned buckets and bits of log, laughing, eating, shouting, their cosy togetherness lit by the glow.

She and Rupert passed by on the outside, like ghosts. Like children gazing hungrily through the windows of a sweet-shop.

'What shall we do?' he murmured.

'Let's just walk,' she said, and they did, trying to look as though they knew where they were going, as though they went to places like this all the time.

'Es, Es, whizz, acid,' a man muttered at them from the shadows, the first person who had talked to them.

They went straight past, keeping their gazes ahead. Then, when he was safely out of earshot, Rupert whispered, 'Do you think we should? Get something?'

And she said, 'Do you want to?'

And he said, 'Yes.'

'What?'

'I don't know. Speed?' Rupert sounded uncertain. Neither of them had done drugs. Drugs didn't exist in Henfield – at least, they hadn't before These People arrived. They knew about ecstasy only what they read in the newspaper, that people died from it, and about acid, that it made you see lobsters. Speed just, well, sped you up. Didn't it? It seemed like the friendliest, the least frightening. 'I'll wait till the next person offers,' said Rupert, boldly. 'Then I'll get some.'

The next person was a man standing outside the sound-system marquee. Lily went inside while Rupert talked to him. There were two stacks of speakers the size of a small house at either side of the tent, and a gaggle of people dancing in the middle. She watched them carefully, so she could learn. They moved differently – some jerky and articulated, others fluid – but they had all merged their bodies completely with the beat; they were focused, in the flow.

A skinny white guy in a baseball cap was nodding along behind the decks, with a rasta MC beside him. The beat banged out hard and fast, over a twisting harmonic sound, like a didgeridoo.

'Forward the revolution,' sang the MC. 'You might stop the party but you can't stop the future . . .'

She was just getting the hang of the dancing when

Rupert reappeared. 'He didn't have any speed,' he shouted, into her ear, 'so I got these instead.'

In the palm of his hand were two tiny squares of paper. Each had a red-and-white mushroom with a cartoon face printed on it.

'What are they?'

'Search me.' Rupert ripped off one square and put it on his tongue. He gave her a gleeful smile. 'We'll find out.'

They had to leave the tent when Lily realised that she was disappearing into a black-and-white spiral wall-hanging. Arm in arm, they wandered unsteadily through the encampment, colours blurring and woozing. Everything seeming to bleed into everything else. The boundaries of Lily's body had turned porous: she soaked everything in, letting it change and shape her.

'Cosmic chanting,' called a woman sitting outside a giant tepee. 'Lullabies for the earth goddess.' She wore a turban and big hoop earrings. Lily stopped.

'Take off your shoes,' said the woman, lifting a large flap of hide to reveal the tepee's entrance and gesturing them in. 'Walk clockwise.'

'What are you doing?' said Rupert, but Lily was already slipping off her trainers, ducking down and stepping through.

Beneath her socked feet was springy rush matting. It was dark, even though a fire was burning in the centre, smoke rising straight up to an opening at the top. She could just make out people sitting cross-legged, the faces dimly lit. Somebody was drumming. The air was pungent.

'Lily!' Rupert hissed after her. 'Why are we in here?'

'I need to lie down.'

They picked their way carefully – clockwise – around the circle of people, and found a spot next to the big drum. The drummer, sitting astride his instrument, was a man with a head like an Easter Island statue. As he played he closed his eyes, swaying. He paused for a moment and started to chant in a nasal voice: *We all fly like eagles, flying so high, circling the universe, on wings of pure light.*

Lily lay back on the soft rushes, her head beside the fire, Rupert beside her. She watched the smoke grow and reach and swirl up into the night sky. The people around her had joined in with the chant, their voices gathering and rising.

Oh-oh, witch-ee-chaio, o-hiyo. Oh-oh, witch-ee-chaio, o-hiyo.

She felt a tap on her shoulder. The drumming man was holding his hands towards her, offering her something. Sitting up, she took it. It was an upright pipe, cool and heavy. She had no idea what to do with it.

'Look,' he said. 'Like this.'

Taking it back, he cupped one hand around the bottom of the pipe and put a lighter to the top. As he inhaled, the top of the pipe crackled and smouldered. Then he gave it to her again, and held the lighter while she cupped the pipe. There was a crackle and she breathed in, a whole lungful. As she breathed out, Lily-aged-sixteen-and-a-bit felt a small circular piece of the top of her head spin off and rise, rise, rise, until it disappeared for ever into the vast star-flecked sky.

She lay down next to Rupert and did nothing for a while. Then she said, 'You know who would have loved this?' He shook his head. 'Father.'

'Oh, Lil.' He put his arm around her, pulled her tight into his bony collarbone. 'Perhaps he's not so far away.'

She was in Rupert's arms, but also back at home, years ago, on the old velvet sofa where Father used to read to her. She was feeling his scratchy jumper against her cheek, the intense calm of being enfolded in his arms. He was rocking her, rocking her, on a deep, black and mysterious sea, and she knew she didn't have to go back, ever, to the house where she and Mother skirted one another in wary, grief-filled silence. To that awful tick-tocking quiet, which pressed down on her like a megalithic stone. In that moment, she knew for absolute certain that if she went home she would lose the power to breathe.

She put a hand to her face, and felt that it was wet. 'You know what we have to do, Rupes, don't you?' she said. 'We have to live like this all the time. We have to run away.'

6.

'WELL, that's the beginning, anyway.' They had finished their drinks, and Dan had got new ones.

Her eyes glittered in the warm light. A flush had crept up her neck, rising to her cheeks. She felt a quiet excitement at being with him, because of the quality of his interest. She turned towards it, a flower towards light.

'You made your escape,' he said.

'Never went back.' She smiled, raised her glass to her lips. They looked at each other for quite a long time then, and it was like a dare, how long they could hold it.

Then he tilted his head, gazed at a point on the side of her neck. 'Tell me one more thing,' he asked. 'What's that?'

Her hand flew up to the five small bruises – one for each finger. They were so faint now, almost completely faded, and she had taken off her scarf in the warm fug of the pub, forgetting all about them. Or, she thought with a guilty start, had she forgotten? Had some part of her wanted him to see? Nobody else had noticed, not Rev, not Aoife, not any of the people who knew her best.

She bit her lip, said nothing. All her confident words had shrivelled up inside.

'That's a hand mark,' he said, in his gentle, rounded voice. 'Who's been putting his hands on you, now?'

She coughed, or made some noise that wasn't really a cough.

'Your boyfriend, was it?'

He had caught her out. It took her a while to find any words. 'He's not my boyfriend any more.'

'I hope not.'

During the next silence she considered the fact of this stranger sitting in front of her, his broadness, his solidity. Perhaps she was surprised that such a big man would be so observant, would zone in immediately on the most sensitive, most delicate thing . . . Outside the window a car passed, its headlights sweeping his face. Dan reached across the table. He took her hand and stroked the back of it with his thumb. It was a funny feeling, his touch. It made things inside her move.

'I know what it's like,' he said. He stopped. He took a sip of his pint. 'I grew up with all that in my home. My dad was awful to my stepmum. However it feels right now, it's not your fault.'

She kept breathing. Didn't look away, or talk, or move her face. Then leaned forward and rested her head on the back of her hands, so he could see only the curly back of her hair. He put his hand between her shoulders, just rested it lightly, kept it there.

DI Wells: Morning, Daniel.

UCO122: Morning, sir – Martin. Sorry. I'm still not used to it.

DI Wells: There's a lot to get used to.

UCO122: You can say that again. Feels like I'm coming up for air.

DI Wells: I remember that, from my tour. Decompression, that's what these debriefs are about, really. So. How goes it?

UCO122: Well. I've made a start. Went to my second meeting yesterday.

DI Wells: And the legend is standing up?

UCO122: I haven't gone into much detail, obviously. I've established that I'm new to activism. That I've had something of a shady past, now making good. Told one of them a bit of backstory about my dad, the miner's strike, et cetera.

DI Wells: How did that go down?

UCO122: Good, I think. I felt confident.

DI Wells: You're drawing from some personal experience there, I believe.

UCO122: That's right, sir. Martin. Although as a child I was on the other side of the lines, as it were.

DI Wells: Ah, yes. Your father was one of us.

UCO122: He was. And being a policeman in a mining community in the eighties, let me tell you, I took some flak at school.

DI Wells: I bet. Must have been tough for you. And for your old man, of course.

UCO122: It was hard, yes. He got death threats – although he never told me that at the time.

DI Wells: I'm sure he's proud of you following in his footsteps.

UCO122: He's a hard act to follow, a copper of the old school, my dad. Always believed in the job, in the importance of doing it well, fairly – even when it's not easy. 'Blessed be the peacemakers' – that was his motto.

DI Wells: Not always a very peaceful business, making peace.

UCO122: I learned that young.

DI Wells: Very good. Now. On to business. Tell me about your wearies.

UCO122: They're a colourful bunch. Not entirely a serious bunch, I would say.

DI Wells: Indeed. But, as you know, they succeeded

in causing some very serious trouble last year. The anti-M11-link-road campaign was an absolute mess. I've never seen chaos like the eviction of that squat – where was it now?

UCO122: Harfield Road.

DI Wells: That's the one. There were hundreds of them. It took us nearly a week to clear. It was all over the news . . . The high-ups are very keen to prevent that kind of situation arising again. Hence your deployment. So I take it you've met – what's his name? Rupert?

UCO122: Delamere. Known as Rev. Yes, he's the ring-leader.

DI Wells: Son of a peer. Stands to inherit a title, and a stately pile in West Sussex.

UCO122: Currently living in a council flat.

DI Wells: Ha! You couldn't make it up. Any other particular individuals we should know about?

UCO122: There is Lilian May McCoy. Known as Skylark. Childhood friend of Rupert, also involved in Harfield Road.

DI Wells: I think I may have seen you with her yesterday, when I was on back-up. In the patrol car, on

Euston Road. I passed you and a female IC1 with blonde curly hair.

UCO122: That's her.

DI Wells: Very good. She's got nice . . .[PAUSE]

UCO122: She does. Yes. That's the one.

DI Wells: I'm sure you'll be keeping a close eye.

UCO122: You can rely on me.

DI Wells: [LAUGHS] And their plans?

UCO122: Their main focus is a big action for the summer. They want to take over a motorway.

DI Wells: They're nothing if not ambitious. Dare I ask why?

UCO122: I'm told they were inspired by the Situationists, sir.

DI Wells: Come again?

UCO122: They don't believe that lecturing people about political change does any good. They aim to create situations in which people can actually experience what life in an alternative society might feel like.

DI Wells: Right-oh.

UCO122: So those who go to their street parties realise

they have the power to take back control of public space – they understand what that means, as it were, emotionally as well as intellectually. They get inspired.

DI Wells: All a bit over my head, that. But write it all up, won't you? And, Daniel?

UCO122: Yes, Martin?

DI Wells: Are you planning to go home, to see the family? You must have been missing your babies.

UCO122: Yeah. Not missing the crying much. I'm going tomorrow.

DI Wells: They're out in Spain, correct?

UCO122: That's right. It's what the wife wanted, while I'm doing this. Her family is there. But it makes visiting a little . . . tricky.

DI Wells: Okay. Well, it all sounds like things are coming along just as they should. Well done, Daniel. I know it's hard at first.

UCO122: So far it hasn't been bad. Fun, actually. Makes a change from real life. Home since the twins arrived is . . . Well. It's hard work.

DI Wells: Whisper it, but that's what they all say – real life is the hard part.

UCO122: Just don't tell the wife that.

DI Wells: [LAUGHS] Don't worry, Daniel. Your secret's safe with me.

7.

THEIR weekly sneakings-off to the pub did not go unnoticed within the world-changing group. Eyebrows were raised, comments made, sniggers sniggered.

'So,' Rev said one afternoon, having come round to help her make a start on the decorating, 'You and Dan, Dan and you, heywhaddyasay?'

'Sorry, what is your question?' she replied sniffily, focusing very intently on the wall, from which she was scraping a fragment of flock wallpaper. She was irritated with Rev, who had turned up to the painting session wearing a fancy black frock coat and black shiny pointed boots, which didn't indicate the kind of serious decorating intent she had hoped for.

'You seem to be getting on famously,' Rev went on, kicking off his impractical boots and stretching out on the sofa. 'Drinking drinks together, et cetera, or so I gather.'

'Thought you were here to help,' she retorted, 'but is it lowdown gossip you're really after?'

'I am *always* here to help,' he said, picking up a scraper and holding it delicately in his long pale fingers. For a moment he appeared to be considering using it. Meanwhile she kept looking at the wall, while also blushing, which Rev clocked, straight off. He practically licked his lips.

'Skylark McCoy!' he exclaimed sumptuously. 'Well, well.'

She held her breath, determined not to rise to the bait.

'And how about Mikey, dare I ask?'

'Mikey's gone.' She ripped at the wall, sending a shower of wallpaper and plaster cascading onto the manky council-issue carpet tiles.

Rev put a hand on her shoulder. 'Really? I'm sorry.'

'No,' she said. 'Don't be. You were right. Enough was enough.' Rev had been telling her to end it for a long time. *Why do you think you can rescue him?* he had said, which had maddened her at the time. What was she supposed to do? Just let him go under?

'I don't care about right,' he said gently. 'Are you okay?'

'Yeah. I'm fine.' She reached for the scraper and held it out to him. 'Get on with it, would you?'

Rev sighed, removed his frilly coat and climbed to the top of a stepladder, where he began scraping away with some diligence.

She relented a little. 'Okay, I tell you what,' she said. 'Dan's got more to him than you think.'

'He's a funny guy,' Rev mused, from the top of the ladder. 'I'm not sure I've quite got his number.'

'Oh?' she said, and for no particular reason her throat tightened.

'When he turned up to that first meeting, he had a Walkman on, playing really loud. Do you know what he was listening to?' Rev paused for dramatic effect. 'Bon Jovi.'

She burst into laughter, out of some weird relief.

'I don't believe it was ironic,' said Rev. 'It's unusual for our meetings to attract somebody of that musical persuasion. Our membership usually errs more towards techno, reggae and trance.'

'Occasionally punk and grunge,' she added.

'Oh, yes, punk and grunge, naturally,' acknowledged Rev. 'Even folk, if you count Ken and Len. But Bon Jovi, I mean really.'

'Rev,' she said sharply, 'stop being such a snob.'

That shut him up. Rev *could* be a snob, not towards poor people, or common people, or whatever the usual snobbery would be, quite the reverse. He directed his disdain towards people he considered to be dull, ordinary, conventional. People he considered to be *too Henfield*. But he got very cross when she told him so. Rev's deep and secret poshness was his Achilles heel: any perceived reference to it touched his sensitive core.

After what was not quite ten minutes of being helpful, he climbed down the ladder, settled himself on the sofa and produced a large pouch of Thai stick from the coat's frilly pocket.

'There's your project, Sky,' he said. 'Slip your new friend a pill or two, and lead him into the light of the rave era.'

It was kind of a joke. But also, not.

8.

THE first time Dan called her she was up a ladder, and the phone rang at least six times before she got to it.

'Oh, you are there,' he said. 'I was about to give up.'

'Sorry,' she said. 'I was decorating.'

'Still?'

'I'll be decorating for ever.'

'Sounds like you need a hand.'

'I warn you, don't say that if you don't mean it.'

'I mean it,' he said. 'Any time.'

She pressed the cold heavy plastic receiver against her ear and fiddled with the curly wire, imagining its route through the wall and under the pavements, through the muck and grime and clay of London's earth all the way from her hand to his hand, from her mouth to his mouth.

'So shall I come over?'

'I mean, yes. Please. But not now. I've got plans.' As it was the first Saturday of April, she would be attending her regular rave in Charing Cross with other members of the world-changing group, including Rev, Mouse and Bendy Aoife. Rev was giving a talk, and they had a guest list.

'People talk, do they, at raves?'

She explained that they weren't going to a nightclub, not some cheesy bar dive where ordinary capitalists and mindless consumers went to drink alcohol and dance the Macarena. This was a rave with purpose, the alternative in action, a dress rehearsal for the better world. At Return to the Source, for such was the rave's name, in between raving you could hear a real-live Native American chief speak about the destruction of his ancestral lands, or have a Reiki healing massage, or join the hunt saboteurs. Rev would be delivering a brief lecture about the evils of consumerist car culture and the possibilities of creative resistance, to an audience who were so high on happy-drugs that they would be really open to new ideas, if only they could remember them the next morning.

'Right,' said Dan, still sounding a little confused, 'so this is the party-as-a-portal thing. The arrow of hope . . .'

'Exactly.'

'Just so you know, I don't do drugs,' he said.

'I respect that about you,' she lied.

Outside the door, a queue of people waited in the cold spring night. They were dressed in fluoro, hair in dreads, plaits, spikes, in mohawks and pixie-buns. They were excitable, hopping up and down, chatting to strangers, passing crumpled reefers down the line. Saturday night was when everything changed, when the trials of the world, the mundanity of survival, the complications of family, the pressures of career, money, housing all fell away and a hazy magical realm opened up and welcomed them in.

As these queuers waited in a state of excitement, suspense, anticipation, they were exchanging cash for chunky white pills with happy faces, doves or Mitsubishi signs on. They were wondering whether to do said pills now in order to avoid whatever door security there probably wasn't, deciding fuck it why not let's get started, washing the first half down with a glug of bottled water, feeling the back-of-the-throat bitterness, the sicky buzz making its way down into the digestive system. The happy-drugs nestled in their stomachs, sending little thrills up their spines, prompting their jaws to move, at first subtly, but gradually, as the hours rolled by, building up to a full-faced eye-rolling gurn. They had prepared for this: pockets packed with chewing gum and sticks of Vicks Vaporub, trainers because obviously, bottles of water, puffas for the cold morning bus ride home.

Skylark and Aoife and Mouse and Rev passed

through the portal, offering their hands to the red-devil-horned doorwoman to stamp, through the entry zone, which was UV-lit and made teeth, dandruff, bra straps glow, and made faces look both luminous and haggard. They emerged into the market, a corridor lined with stalls offering hemp-cloth bags, flowing printed skirts, skull-shaped bongs, fractals, lighters, more bottled water, homemade 'special' brownies, CDs featuring the wisdom of Ram Dass, glow-in-the-dark bongos, gaily painted didgeridoos, rasta hats, string vests and nose-piercing.

A door on the left was signposted 'The Well', with a blackboard propped up advertising the talks schedule:

10.30–11 p.m. Mystic Circles: the Cerealogist Quarterly on the latest developments from the field(s).

11.30–midnight Radical Currents: learn the ancient art of water-divining.

00.30–1 a.m. The Road to Resistance: taking back the streets, a tutorial with the Reverend.

'Arseholes,' said Rev. 'They've put me on last. How is anyone supposed to make sense at that time of night?'

'Never fear, Revvy,' Aoife reassured him, putting a lithe, muscular arm around his shoulders. 'It's not like anyone will be listening.'

After a brief foray to the bar for cider, they headed first to the chill-out area, to chat, drink and wait to come up. This was a small room in which many sticks of burning incense could not quite disguise a penetrating smell of damp. The floor was covered with cushions and rugs, and on the walls hung black cloths and various papier-mâché UV extraterrestrials. The room was dominated by a huge fluoro mushroom growing up one wall, its stem the size of a small tree trunk.

'Another night at the office,' said Rev, holding up his plastic cup in salutation.

Skylark tapped it with her own, looking sneakily over his shoulder just in case she should spot Dan. She tried to imagine what he would make of all this: his jury's-out slanty smile, his slightly-too-neat jeans. And in doing so she was beginning to see things as he would see them: this habitat that had been so naturally hers appeared all of a sudden just a little foolish and ridiculous. She caught herself momentarily thinking, Why, after all, have I chosen to wear a skintight silver PVC dress over tie-dyed rainbow leggings? Couldn't I maybe have worn something more normal? As though she was developing her own internal Dan.

'Is anyone feeling anything?' she enquired, once they had sat down.

'A low-level buzz,' said Rev. 'I might take another half. Just to be sure.'

'I feel a bit sick,' said Aoife.

'Same,' Skylark said, although she didn't know whether that was the drugs or perhaps the arms-related anticipation.

Mouse said nothing. His bolt-upright body was vibrating slightly. Perhaps he was on the brink of Nirvana.

'I'm going to walk,' she said, standing up. The room whirled, and shivers radiated out from her stomach. Yes, definitely, the happy-drugs were kicking in already. 'See y'all on the other side.'

Keeping one hand on the wall, she picked her way over the batik textiles of the chill-out zone and emerged from the cloud of nag champa into the darkened hall. Through the double doors at the end was the main dance-floor, where the air was hung with dry ice, and green fingers of laser reached up and down, tracing the outlines of a mass of moving bodies swaying and waiting for the beat to drop. The tune was the one that was everywhere at that time, the one with the beat like clanging bin lids and the insistent robotic vocal:

> You've got to believe in something
> You've got to believe in something
> Why not believe in me?

And with uncanny or even fateful timing, the moment the bass dropped she looked up and there he was, standing in front of her. Their eyes locked immediately and some electrical thing flowed between them. She sensed it even more strongly than she had before, her sensors enhanced by the happy-drugs. She had a thought of uncanny clarity, like a memory of something that hadn't happened yet: how his warm dry minty-fresh lips would feel against hers. She wanted to say something outrageous, like *I want you*, softly into his earhole, *please*, and feel him tremble.

His head and his hair were outlined in green by the laser. He mimed *Hello* against the rumble of bass, then pointed at the door and panted in the universally acknowledged sign for *I need an immediate drink*.

In the bar area the floor was wet and stank of sour apples. By the door there was a leatherette couch in which two crusties were sitting, locked in intense conversation. One had a rattail, and the other a top hat.

'You know, they've developed new surveillance techniques,' the rattail was saying. 'They can control your brain via a tiny bone at the back of your nose.'

'A double vodka, please,' Dan said to the barman, a note of real urgency in his voice. 'And a cider.'

'Just a water, actually,' she corrected him too late, and he raised an eyebrow.

'Oh,' he said. 'It's like that, is it?'

The drinks arrived and he knocked back the vodka and kept the cider for himself. She took a long pull of ice-cold water, not because she was thirsty, just because she should. Hydration was key. Dan had been growing his beard, and there was something else different about him. He almost looked like somebody who would usually be here.

'What have you done?' she asked, trying to work it out. 'You've done something.'

'Oh,' he said, embarrassed, 'you mean this?'

He lifted his hair to show her the sides and back, which had been shaved into an undercut.

'Aha! See? You're becoming one of us!' He pulled a mock-horrified face. 'So what do you think of your first rave?'

Now his grimace was real. 'Ask me after a drink.'

'I've got something else that might help.'

She shook another pill out of the baggy in her purse and bit half off for herself, washing it down with the water, then offered the other half to him.

He gave her his slanty look. 'I told you, I don't do that.'

But she held it there between them, glowing UV-whitely. He held her gaze for a long moment, like he was at the corner of a street and couldn't decide which way to turn.

'What?' she said, slapping him lightly on the wrist. 'Stop thinking.'

He took the pill from her palm, tipped it into his mouth and swallowed. 'I already have,' he said. 'Clearly.'

Dan came up like a door opening, and light pouring through. He came up like a sunrise, like the dawn of a new day. The slantiness melted from his face, to reveal his open, generous, childlike nature, the nature that we all have hidden soft and pulsing beneath our shells and spikes, our scales and claws. He wouldn't let go of her hand. They reclined together on the beanbags beneath the giant mushroom.

'Sky,' he kept saying, 'I've never felt like this.'

She stroked his hair. 'You mean the love, Danny boy?'

'There's so much of it,' he said wonderingly, his bug-eyes rolling. They'd lost the others. Even Mouse had disappeared – levitated off to another astral plane, perhaps. 'It's like I can't fit it all inside, it's spilling out. I love everyone in this room.'

'Why stop there?' She egged him on. 'See if you can love everyone in the world.'

'I can,' said Dan, grasping her hand. His eyes were darker than ever, irises just a thin border around each pupil. 'It's easy.'

But now a tetchiness was rising in her limbs, a pins-and-needles itch, rising to a rushing sensation so intense that if she didn't move soon she might neglect to breathe. 'We need to dance,' she told him. 'Come on, get up.'

'I like it here.'

'Listen to me. We have to keep moving,' she said sternly, standing up and pulling him upright. 'Or the love zone can turn into the mong zone.'

'Yeah,' said Dan, bouncing on his heels, shaking out his arms, 'you're right. Moving is good. Ooh, I'm on the edge. It's speeding up. This might be too much.'

'It's fine. Have some water,' she said, pressing a bottle into his hands. 'Keep drinking. Let's go.'

They found Rev and Aoife by the speakers, absorbing the heavy bass vibes through their feet and their chests. Rev was doing make-a-box robot dancing, and Aoife had a more flowy goddess style. Sky closed her eyes and felt the beat move through her body. Sometimes it took her a while to tune in to her dancing self. It started in the shoulders, then moved to the hips. When she got into the flow, she could dance all night. She sometimes danced in the bus, all the way home. Dan was standing stock still, watching the lasers. He lifted up his hand wonderingly, trying to catch the green beam of light. Then, cautiously, like an explorer setting foot on uncharted terrain, he began to move.

* * *

'Everything's all shimmery,' he said. 'Can you see it?'

At half five in the morning, they were walking, arm in arm, heads still in the fluffy pink clouds, bodies half dancing, down Embankment. A milk float trundled past; birds were starting to sing in the giant plane trees.

'You're coming down,' she said. 'I love the comedown. It's my favourite bit.'

He stopped, leaned against the wall between the pavement and the river, tugged her towards him. She could see her own face in his too-big pupils, and feel his weight. And at this junction between one thing and another, a voice spoke in her mind. It said: *I want you in my life.* It was a very strong, firm instruction, perhaps even a command. She had never heard this voice ever before, but it reverberated through her skull and her body, a string plucked in her heart.

His lips were soft at first, then quickly ravenous, and she was ravenous too, and they were gulping and clinging to each other, their bodies coming together like two halves of something. He tasted just as freshly minty as she was expecting and also of something else, like earth.

'We could go back to my flat,' she said, coming up for air.

He kissed her again, gentler this time. 'And the boyfriend?'

'I told you. He's not my boyfriend any more.'

'Just checking,' he said, kissing, kissing.

In her bedroom, he spent ages messing about with the curtains. She didn't immediately wonder why, as she was too busy admiring the following things about his bare torso: (a) how broad his shoulders were, (b) the dimple on his shoulder blades, which created a kind of rounded hill and then a valley that traced the edge of the blade, some really interesting and complex cartography that she would have liked to try drawing, or something, and then, (c), and this seemed quite a lot more urgent, the way his waist sloped smoothly down into his jeans, which were still on.

'Oy,' she called from the bed, in an unladylike manner, 'stop fucking about.'

'Coming.'

He crossed the room and flicked the light switch, plunging the room into black. For a moment, she couldn't see anything at all. The darkness was solid, like a wall.

'What the—'

And then, before she could even try to make out any of the familiar landmarks of her bedroom – the giant heap of clothes, the hatstand with her anatomically correct monkey suit hanging on it, the Tofu Love Frogs poster – he was on top of her. He was running his hands up inside her dress and he was pressing his mouth into her neck. The weight of him was immobilising. She still couldn't see. She wanted to see. She didn't care.

'Why did you make it so dark?' She reached out a fumbling hand and turned on the bedside light.

He winced, shading his eyes, then grabbed the switch and turned it off again. 'It's better like this.'

'But don't you want to—'

He put one finger over her lips. 'Honestly. Try it.'

He shifted his weight again, pressing her down hard, and she lost any will to argue. She took Dan's plump earlobe between her teeth, bit it gently, and felt him shiver.

He whispered tenderly, 'It's so I'm not blinded by your tights.'

She giggled, then whispered back, 'I've already taken them off. Here, feel.'

Turned out he was right about the dark. She could still feel him, trace those contours with her fingers. But in that pitch-black room she didn't have to worry about being her ordinary everyday self: she could let herself be something and somewhere else. 'Be strong,' she whispered urgently. 'I don't care if it hurts.' He pinned her down. He forced her head onto the pillow. She could only just breathe, poised in some still place between pleasure and pain.

'Like that?'

She said, or maybe just felt: 'Yes.'

She was lowering herself into a river, feeling the current pull at her, knowing that it would sweep her away. The thing was not to resist: it was to let him take control completely, to let go. And, sure enough, he took her right out into that deep dark sea, to a place in which she no longer knew where she ended and he began.

9.

DAYS passed, and despite the fresh-minted spring sunshine she didn't feel like going out. She couldn't even face more decorating, sat there in the long lonely evenings after work feeling overly conscious of the telephone, which stubbornly slumbered away in the corner of the bedroom. It woke briefly once, but instead of the deep buttery tones she had hoped for, she got a bracing earhole-cleanser of crisp Queen's English.

'Broad beans,' announced Mother. 'Would you like some?'

Mother was calling from a parallel universe, a.k.a. Henfield. She could picture the scene exactly: Mother would be standing in the square boxy living room, next to the mantelpiece with the square tick-tocking brass carriage clock on it, and the baby pictures, one on either side. She would be wearing her slacks and her cardy. Her bifocals would be hanging around her neck. The feelings that this image evoked in her would once have been pure irritation, but nowadays inclined towards tenderness, nostalgia or even longing. She wanted to find her way back to Mother, after all these years, find some route across the universes.

Static crackled. The line from Henfield was always, for some reason, awful. 'They're delicious in a salad. I've already given them to all the neighbours.'

These days Mother spent most of her time at her allotment, which took the edge off. After many years of complete mutual incomprehension, mother and daughter had figured out they could communicate via vegetables.

'I don't know much about the broad bean,' Skylark replied, in an upbeat tone designed to disguise her disappointment that it wasn't Dan calling. 'What does one do with it?'

'Blanch it,' said Mother. 'Or you can actually make hummus.'

'Hummus!' she exclaimed wonderingly.

'Yes! Apparently it doesn't have to be chickpeas. I cut the recipe out of the paper.'

Poor Mother. Her daughter – who didn't even call herself Lilian anymore – had disappointed her in so many ways: failing to go to university, rarely if ever brushing her hair, living in a succession of camps and squats rather than beavering away like a hardworking squirrel hoarding her nuts for a rainy day. Mother would have loved her to be a lawyer, a doctor, a *system* person, someone she could have boasted about to her respectable Henfield friends.

'If you're so into world-changing,' Mother would say, whenever she phoned from the camps to reassure her that she wasn't dead or banged up yet, 'you should study politics at university. You could get an internship at Westminster and become an MP.'

But her daughter had zero interest in that kind of politics, or in that kind of life. She was busy living the world she wanted to see. Her kind of studying happened in makeshift tarp-and-hazel benders on verges and cargo nets in forests. She studied the riot police and bailiffs, the defenders of those corporate capitalist interests, as they moved into each beautiful world-changing encampment with their plastic shields and their body armour, their dead eyes and hatchet-faces, emissaries of the system, of power, of death. She studied the bulldozers and the chainsaws as they wreaked damage that could never be undone.

She studied all those things and she drew her own conclusions.

'Anyway,' said Mother, over the crackle and hiss, 'I'll give you some, if you come over for lunch this weekend.'

'I might have plans,' Skylark replied. When she was younger, she had often thought that *The wrong ones died*.

One from each team. Because Janie, pinky plumpy Janie, who had gone in her sleep when she was two days old, would always be Mother's darling. While Father, quite clearly, was where Skylark got her inner landscape from. He had had a scent of the forest on his breath; had written poems at his desk at McCoy Insurance Limited and studied crop circles in his spare time, their formations, significance, energy fields.

He had taken her with him, once. The circles had always fascinated him, but the obsession took hold after he got the diagnosis. He subscribed to *Cerealogist Quarterly* and started a correspondence with fellow crop-circle enthusiasts all around the country. As his body wasted and bloated from the chemo and the tumour, he became ever more preoccupied with theories about miniature tornadoes and energy fields, about ancient channels that ran between the circles and sacred sites, like Avebury and Stonehenge.

The riddle of the circles came to represent, for him, some much bigger riddle, about life and death.

During those last painful months, his cerealogist friend Graham would phone him when a new one appeared, and he would haul himself out of bed, hobble out to the car and drive off to see it. Mother would roll her eyes but, then, she had always rolled her eyes at Father: when he had given poetry readings at the dinner table; when he had passed his evenings studying Zen calligraphy; when he had spent an entire month's salary on Russian bonds, after the idea had come to him in a dream. 'Your father is not a practical man.' She would sigh, as if this were some terrible affliction, as though it would be better for him to spend his days thinking about Tupperware and parking.

Skylark had no interest in practicality: she wanted love and pain, wanted things she had never tasted. Her inner landscape was towering trees and standing stones, open sky, brambles in her hair and mud beneath her fingernails, and this was a place she knew she shared with Father.

That afternoon, she had followed him down the furrowed tractor-path through the wheat, brushing her hands against the sticky seed-heads. His illness had given him an unsteady lurching gait; he leaned heavily on a cane. She could see him so clearly, even now, in his old brown mac, with his dark, bowl-cut hair. That afternoon it was just the two of them.

'This one is very fresh,' he was saying. 'It appeared last night, according to Graham.'

They emerged into an open space. Underfoot, the wheat stalks were bent neatly over, flattened against the ground in a smooth spiral. It was a perfect circle, and the late summer light divided it into four precise quarters:

two tawny brown, two pale gold. She bent down to look at the flattened wheat.

What was the answer?

As Father struggled off up the track to look for more markings, she stood still, her feet placed exactly on the central point of the spiral. She closed her eyes and sank into the golden space behind her eyelids, tuned herself into the invisible forces moving through the earth. Around her the wheat stalks rushed and whispered, and in the distance, a bird gave a lonely cry.

Could she sense it?

'What are you doing? Funny girl.'

Father's rounded northern vowels summoned her back to the field. He was standing a few feet away, the tape measure in his hand. Behind him, the sky was white as an eye – not a trace of blue, even though it was August. It was a curious nondescript light, so it could have been breakfast time, or evening, or any time in between. He was so sharply defined against this brightness that, for a moment, she felt she was seeing him as a stranger would. Leaning on his cane, his strong body diminished, his eyes squinting, the shadow of pain on his face.

A thought appeared in her mind: *You won't be here long*. For a fragment of a second she understood that completely, and then it was gone.

'Lilian? Lilian? Are you there? I can't hear you . . .

The line is terrible,' Mother was saying. Mother, who had lived; Mother, who was still here, after all these years, trying to reach her, to say something, to fill the gap. 'Listen, don't worry, if you're not in, come next weekend instead, come any time. Okay? Bye, now, darling, bye . . .'

10.

WHEN Tuesday rolled around again Skylark sat chatting to Aoife on the saggy old sofas. Mouse and Rev and Ken and Len nattered and smoked and dunked as per usual, but she kept finding herself staring at the door, willing him to walk through it. The half-motorbike half-stegosaurus in the corner – a creation of Rev's for Glastonbury – eyeballed her, as if it knew something. *What did it mean*, she asked it silently, *the voice that said, I want you in my life?* but the stegosaurus wasn't letting on.

In the interminable time since the naked things had happened between her and Dan, certain sensations and memories had been replaying in her head on a tortuous loop. His weight, his minty-earthiness; the steadiness of his body, how he held her firm, one solid thing in this wobbly old world.

'Right,' said Rev, balancing a notebook on one skinny knee. 'So, let's get started. We've a lot to get through.'

Open the door. She issued this order silently, internally, but with such force that it beamed out through the roof and into the night sky outside, like a bat signal.

The door opened.

Dan slipped quietly into the room and chose a chair at the furthest-away point from where she sat. He caught her eye briefly and gave her a brisk, business-like nod. No smile. He was wearing a checked shirt and black combat trousers, and his new haircut had grown out a bit. The stubble was now a full-fledged beard.

'Ah, our northern friend,' said Rev. 'Way-ey lad. Thought we'd lost you.'

'I'm not Geordie, Rev.' Dan's voice had an edge.

'All right, calm down,' said Rev, in Harry-Enfield Scouse. 'It's just you've been AWOL and we've only got two months to sort this action out. We haven't got much time.'

'I know,' he said. Then he looked straight at her. 'Sorry.'

This came out so awkwardly that there was a rustling around the circle as people shifted in their seats. Her face burned.

'Okay, well, moving on,' said Rev, with an almost imperceptible roll of the eyes. 'Let's talk storage.'

After they finished Skylark picked up her bag and coat and headed straight for the door. She needed to get away and fill her lungs with fresh air, as they didn't seem to be working properly. Outside it was, as Mother would say, mild. A mild May evening. The street looked grubbier than usual under such a clean clear sky.

'Sky!' He was behind her, pulling on his coat. He jogged to catch up, but she walked faster. 'Wait, would you? Hey!'

He caught her arm, stopping her in her tracks. 'I was thinking we could go for a drink.' She shrugged, looked away. 'Or, no, I tell you what. Come here for a minute. I want to show you something.'

After a brief and futile fight with herself she followed him down a small alleyway through a low red-brick estate. At the bottom of the alley there was a green metal gate, which he leaped over easily, then turned back to give her a hand.

'Where are we going?'

'I found this place the other day. Thought you'd like it.'

They were standing on the towpath, a cracked walkway dropping precipitously into the algae-green canal. The air tasted metallic, with a hint of something musty and festering. It was chillier now the spring sun had gone down, and damp. Dan set off at a pace, and she kept up, breathing warmth into her hands, wondering why she had allowed him to derail her plan to walk away in a strong and dignified manner.

'It's just down here. Not far, promise.'

The geometric skeleton of the gasworks loomed from behind a low wall. She held her breath against a strong waft of cidery piss as they passed under a road bridge, footsteps ringing sharply. Canals and towpaths were not leisure destinations then; they weren't yet places for pleasant bike rides, waterside dining or pop-up mocktail bars. They were damp, derelict relics of the industrial age, places where shopping trolleys went to die. Up a slope they came to a lock, two mighty black gates holding up a weight of dull brown water.

On the other side of the lock she saw a small, white cottage. It was so unlikely. It should have been in a fairy-tale forest, she thought, or at the very least in a little village called Something-upon-Something. Not here, in the middle of industrial London, criss-crossed by the gasworks' shadow. She felt somehow sure that if she walked past this very same spot tomorrow the cottage wouldn't be there.

'Isn't it like something from *Alice in Wonderland*?' he said. 'I knew you'd like it. This is the kind of place I want when I'm old. A little house where I can sit and watch the sun set.'

On the far side of the house was a sweet little garden: flowerbeds dotted with cow parsley, an elder holding out white plates of blossom, an apple tree with frilly green leaves. A small wooden rowing boat was moored outside.

Dan stepped onto the top of the lock, then reached back to take her hand.

When they'd clambered in, the boat drifted free of its mooring with a creak, and he fixed the oars into their brass clippings, rowing with a slow splash. She watched him moving before her, tried to be objective, failed. She wanted things from him, wanted him to hold her down.

'So you've been away?'

'Yeah. Went to see my dad.'

'How was that?'

He grimaced. 'It has to be done occasionally. Like eating your greens.'

'Isn't it terrible that we feel like that about our parents?'

'You wouldn't say that if you met him, believe me.'

The water lapped at the boat. Very far away, London beeped and roared, and people shuffled pointlessly between offices and houses and supermarkets.

The dark mouth of a tunnel yawned over his shoulder. He kept rowing, and it swallowed them. She pulled her puffa around her more tightly. The sound of the water rushing over the oars was more tinny in here, echoey. The other end was a faint glow over his shoulder.

'I'm really sorry, Sky,' he said.

'What for?'

'That I didn't call. It wasn't that I didn't want to.'

She kept on breathing in the dark dank air. They

emerged suddenly, too suddenly, from the tunnel and re-entered the ordinary everyday world. Car headlights reflected on the darkening water, drunken shouts, the roar of a motorbike.

'What stopped you, then?'

He paused, put the oars down in the bottom of the boat. 'I don't want to mess you up,' he said.

'Hey, don't say that,' she said. 'You won't.' She reached over and stroked his cheek. He took her hand and gently placed it back by her side.

'You don't understand . . .' and his voice was heavy now '. . . I will.'

There was a pub on the towpath and they stopped for a drink before taking the boat back. It was a funny little place, tiny, on the edge of an estate; the clientele all seemed to know each other. While he got the drinks in – he always got the drinks in, never seemed to worry about money; she would have had trouble admitting how much she liked this about him – she sat outside at a grimy table, watching the sky, which had turned brilliant pink and mauve.

Dan came back with two pints and two tequila slammers, slices of lemon and a salt shaker.

'Woah there, it's Tuesday.'

'I thought you were supposed to be the fun one.'

She took the offered slammer, because she always took an offered slammer. They dabbed lemons between thumb and forefinger, poured on a dash of salt. Lick, gulp, suck: she was a sword swallower. Fingers of heat reached from her stomach up to her chest. Shaking his head at the kick, Dan leaned forward, pointing to her, and then to himself. 'So. Do you think the universe wants this?' he said.

She didn't understand. 'I think the question is: do you?'

He rubbed his eyes. 'Meeting you, doing all this – it's opened me up in ways . . . I can't explain it. The last few weeks have blown my mind.'

'But?'

'I'm confused. I don't know what I want.'

'That's okay,' she said. 'Maybe some day you will.'

Black birds swooped low over the canal. From inside the pub, the lulling low sound of men talking. Their hands were very close on the table; close, but not touching. He moved his away, stuffed them into his pockets. Something ached right in her centre, the familiar place, too deep to touch. She wrapped her arms tightly around her chest, and avoided his eyes.

DI Wells: I need to take some leave, Martin.

UCO122: When?

DI Wells: As soon as possible.

[PAUSE]

DI Wells: Is everything okay, Daniel?

UCO122: I don't know. It's just – it's kind of getting to me, a bit. The job. It's the confusion.

[PAUSE]

I'm fine. Sorry. I just need to go. See the twins. Speak to Adrienne about things. It's been too long.

DI Wells: How long has it been?

UCO122: I went once. Beginning of March. Not since then.

DI Wells: March? That's nearly three months!

UCO122: I know. It's far too long. They won't even recognise me.

DI Wells: Why didn't you go before?

UCO122: It wasn't great, last time I went back. For either of us. Adrienne was just getting used to having the place to herself, dealing with the twins, just her and her folks – and then I turn up and get in the way. I wasn't exactly flavour of the month.

[PAUSE]

But, no, it was my fault. To be honest, I couldn't relax. I just kept thinking: what if someone comes round to my duff address and finds I'm not there? What if they call and I don't answer the phone? How am I going to explain missing a meeting?

DI Wells: That fear is awful. I remember it well from my tour. But you must keep going home. This job is tough on marriages.

[PAUSE]

Without wanting to get too personal, Daniel – how long have you been married?

UCO122: Five years.

DI Wells: You must have been young.

UCO122: We were both twenty-three. So, not really.

DI Wells: Young enough.

UCO122: We met at school, so. Went through a fair bit together, early in life.

DI Wells: Childhood sweethearts. And the twins are a boy and a girl, right?

UCO122: Pedro and Anna. They're nearly a year old, now. Walking. Or so she tells me.

[PAUSE]

> But apart from all that it's great. I've been – enjoying it. This whole experience. It's been a taste of something different.

DI Wells: We often end up feeling that way. When you're in the field, when you experience the camaraderie, and all the rest of it – you realise that these groups really have their place. We come to appreciate some of the good things about them.

UCO122: I'm glad it's not just me.

DI Wells: At the same time, we must remember the impact their activities can have on other people's lives, businesses and property. It's a fine line. Our role is to make sure they're policed appropriately and proportionately.

UCO122: The world needs them, and it needs us. Yin and yang.

DI Wells: [SNORTS] Yin and yang! Dear me. Go home for a few days, Daniel. Have a beer, watch some football. Scrub off the patchouli. I'll put in an application for emergency leave. I'm sure I can get that cleared. Please don't leave it so long, next time.

UCO122: Understood.

DI Wells: Obviously we'll need you back in good time for the motorway protest.

UCO122: Of course. I wouldn't miss it. Thank you, Martin. I appreciate your understanding.

DI Wells: I know it may not feel like it now, but that very intense confusion is actually a sign that you're doing well. You're becoming immersed. The big challenge of the job is living with that. Embracing it, even.

UCO122: I'll try. Thanks again, Martin. I'm sure things will be clearer by the time I get back.

11.

'DAN's there,' said Moll, ending the call and tucking her brick-sized mobile phone back into her belt. 'He's a mile away, Point B. Parked up with Jez and the road crew. They're about to leave. And he's spoken to Ken and Len. The first tubes have departed Liverpool Street.' She checked her watch, and grinned at Aoife and Skylark. 'In about ten minutes, ladies, two thousand people will be arriving at Shepherd's Bush station.'

'Oh, my God,' said Aoife, 'ohmyGodohmyGod. Okay. This is happening.'

She did a quick handstand, to calm her nerves. Skylark was pacing, black notebook clutched in her hand. She had been pacing ever since they'd arrived in their designated position an hour ago. Above them, the sky was blazing July blue, and seemed to press down on the three women with an oppressive sense of now-ness. Around them, the city was holding its breath.

This was the day. The only day that mattered. The day they were going to take the motorway.

Over the five months of its development, the plan had assumed astonishing complexity. It had spilled out of the black notebook, which was now in tatters, collapsing at the spine. The world-changing group had rehearsed every element, until they could execute each one with military precision. At every stage, there was a plan, and a back-up, and a back-up to the back-up. And yet still there was an element of near-overwhelming chaos, a feeling that the whole thing was, always and for ever, on the brink of falling apart.

'Will it work?' they had asked each other over and over, alternately disbelieving and then believing again. 'Could it actually work?'

Moll, Aoife and Skylark were the support crew, and as such their job was to wait in a large, derelict timber barn, a relic of some long-forgotten age when Shepherd's Bush might actually have had some resident shepherds. It was now squeezed incongruously between three tower blocks on one side, and the motorway slip road on the other. In front of them was a large blank concrete wall, from beyond which came the ceaseless, restless whine of cars.

At the exact moment that the first tube train carrying punters arrived at Shepherd's Bush, the road crew would block the northbound side with the staged car-crash. That was when they, the support crew, would prop up a ladder and climb over the wall onto the motorway, ready to meet the vans, unload and assemble the tripods (thanks to many hours of practice, this would take them approximately four minutes thirty-four seconds).

Once the crowd arrived to block the southbound side, it was job done: London's biggest ever world-changing street party would be under way.

There had been some debate around why all the world-changing women had been bunched together in the support crew. It was a purely practical decision, said the world-changing men, nothing at all to do with the dominant patriarchal ideology. When it came down to it, jobs were allocated on the basis of skills. And if it just so happened that the men were more experienced in driving large vehicles, and doing brave/technical things, that wasn't their fault, now, was it?

From very nearby, a police siren wailed. Moll wiped her brow with a greying lacy handkerchief and stuffed it back into her ample bosom. 'The fuckers are all over us,' she said. And it was true: as they'd walked from the station earlier, police vans had been parked up everywhere, crawling over the whole area. 'I don't understand how they knew the location, when only about ten of us do.'

Skylark shrugged. 'It's fine. We're so close now. There's not a lot they can do about it.' But her face didn't quite share that confidence.

Moll's phone rang again. It was Rev, in the other van. 'Shit,' she said. 'Shit.' She turned ashen. 'Right,' she said. 'Okay. I'll tell them.'

She turned to the others. 'There's a police cordon,' she said. 'Right in front of the station. None of the punters can get through. Rev says we're going to have to do it on Shepherd's Bush Green instead.'

'What?' Aoife and Skylark stared at her, disbelieving. Was this when it all fell apart? Was this when they realised that the whole thing was impossible, that the plan had been as stupid and crazy as they'd always, deep down, suspected? Because this much they all knew: if a rave on the motorway was epic poetry, a rave on that scrub of dejected grass outside the station was a crappy limerick.

Aoife was hopping up and down. 'Fucking lunch-outs,' she fumed. 'Why can't they just break through the cordon?'

'We can't give up on the motorway. I'd rather go home,' said Skylark. She had stopped pacing: her mind was working too fast. 'You know what?' she said. 'Call Dan.'

'What's Dan going to do about it?'

'This is no time for sweet nothings, Sky,' said Aoife, sharply.

Skylark glared at her. 'Fuck off, Eefy,' she said, then turned to Moll, holding out her hand. 'Give me the phone.'

He answered after one ring, and the sound of his voice on the line stilled her jolting world. 'Listen, we're not giving up,' she said. 'We have to think of another way.'

'I don't know, Sky,' he said. 'If they've cut off the road, maybe we just have to accept—'

'We're not accepting anything,' she interrupted. 'Not yet. There has to be a way we can get around the police.'

There was a long pause. This was a moment, she would realise later, when he could have gone one way or the other. Another tipping point. Although perhaps he had already tipped. It didn't take him long to decide. 'Okay,' he said. 'Let's see.' There was a rustle on the line. 'I'm just looking at a map. Okay, here's what we tell Rev. There's another way through. They can get around the police cordon, through the backstreets. If enough of them do that, the line will fall apart. The police won't be able to regroup in time.'

She nodded. 'Okay, that's a plan. I'll call him now,' she said. 'And you lot – now is the time to take the northbound side. You hear? Do it now. Don't wait around.'

Again, he hesitated. 'Are you sure?'

'Of course I'm fucking sure, Dan,' she said. 'Just go for it.'

It worked – for five minutes. By the time the support crew scaled the wall, Jez had crashed the old banger, with Dan's van in front. Rev was already parked a few metres away, on the hard shoulder, his vehicle loaded with another sound-system, the sand for the sandpit, the kitchen gear, the fluorescent alien heads, and the entire Belgian circus troupe.

There was a furious beeping as the entire northbound side of the M41 stopped moving. The southbound was already empty, rerouted around the cordon. For a few seconds, Skylark, Aoife and Moll stood there together, dumbstruck, witnessing this extraordinary thing stretching out before their eyes: a barren wasteland of road, completely empty of traffic. A concrete desert. A motorway, with no cars.

But there was no time to stand and stare. They had four minutes thirty-four seconds to get the tripods up. Rev raced over to help them. Jez was out of the car, looking dazed. He had taken on the scary job, the one nobody wanted – crashing deliberately across two lanes of a motorway: no joke. Rev clapped him on the back. 'Good job,' he said, and Jez glowed.

The tripods were up, Jez and Mouse sitting at the top, by the time the first police motorbikes wound their way past the roadblock. Skylark and Rev managed to scramble on top of the van, but the officers got Dan straight away, pushing him down roughly onto the tarmac and jamming his hands into cuffs. Moll was next, then Aoife, then the Belgian circus troupe, one by one. Sunflowers blew in the wind, lentils scattered, banners fluttered uselessly in the wind. Again they were on the brink of defeat and devastation.

But, beside her, Rev was getting to his feet.

'Sky,' he said, and his face was lit up like a bulb. 'Over there, look.'

She followed his gaze, standing up and squinting into the sun. In the grey flat distance, things were moving. First one person came running up the southbound side, then a handful, and then, in one pulse-racing rush, there were hundreds, thousands. The crowd from the tube station had broken through the police line. It was a tidal wave of people, hair a-flying, feet a-pounding, bongos a-drumming. They were shouting up, calling, dancing to nothing, begging for music, for the world-changing, hope-firing party they'd been promised.

The voice of one of the officers down below was just audible above the noise.

'Orders are to regroup,' he said. 'We're to de-arrest them all, and retreat.' The other officers stared at him, until his face contorted into a shout. 'Take off their cuffs, and let's get out of here. Now!'

As they beat their retreat, the sound-system kicked in, with a great roar of bass and a mocking voice:

What we are talking about here is a total lack of respect for the law . . .

How to describe the biggest illegal street party London had ever seen? The glorious, never-to-come-again afternoon when the barren grey wasteland of motorway, that tarmac desert, bloomed into riotous, beautiful, smiling, raving life . . .

A portal

An arrow of hope, shot into the heart of a dying city

It was everything the world-changers had dreamed of. The universe apparently wanted what they had to offer, after all.

Aoife moved elegantly through the crowd on stilts, grey wig piled on her head, Marie Antoinette dress creating a giant tent beneath which people took turns in drilling up the road, planting trees on the tarmac.

Children played in the central reservation, building castles out of sand. The Belgian circus troupe depicted the downfall of capitalism by means of expressive movement. It all happened, every bit of it.

And while it was happening, at a certain moment, Skylark and Dan sat down next to each other on a sofa somebody had dragged into the road. They were witnesses, now: the party had its own life; it no longer needed them. A group of clowns were waddling about in front of a line of police, wearing big shoes and red noses, sticking flowers into epaulettes. The buzz of earlier had faded into a happy, peaceful glow. Dan was nursing one arm, injured during his arrest, but he put the other around her, and she leaned into him, rested her head on his shoulder.

'Thank you,' he said.

'What for?'

'For all this—' He stopped, then added, his voice rough around the edges, 'For completely changing my life.'

'I didn't do anything. That was all you.' She wasn't looking at him: she rose and fell with his breath.

They sat for a while longer, still and warm and together. 'Oh, and,' he said eventually, 'I think I know, now.'

'You do?' She wrapped her arms around him, more tightly, and he kissed the top of her head.

'I do,' he said.

DI Wells: What happened to your arm?

UCO122: It was – I was arrested.

DI Wells: This was the initial arrest, before the crowd broke through the line?

UCO122: That's right. The arresting officer was a total thug, Martin. The kind of guy who makes you ashamed—

DI Wells: Okay, calm down, please. Just tell me what happened.

UCO122: He twisted my arm behind my back. Nearly dislocated my fucking shoulder.

DI Wells: I'm sorry to hear that. We can make sure you receive medical—

UCO122: There was no way this was reasonable force. And I wasn't the only one. Guys like that shouldn't be posted to these events. He was just looking for an excuse—

DI Wells: I do understand your point, Daniel, and I'll take that back to their supervising officer. To be fair, he wasn't to know who you were. Make sure you get yourself to the medic this afternoon.

[PAUSE]

Can we move on?

UCO122: I guess.

DI Wells: What is your assessment of Saturday?

UCO122: Well. It was an amazing day. A triumph.

DI Wells: How so?

UCO122: I mean, just take a look at the papers: 'Chaos as 500 Join Motorway Street Party'. 'Tens of thousands of pounds of damage to national infrastructure'. We totally underestimated the numbers. I personally didn't think that we – that they would attract so many people to the event. You – we – would have needed more boots on the ground to keep things under control.

DI Wells: Don't be hard on yourself, though. You did everything we wanted you to do. As a direct result of your intelligence we were able to put in place the line outside the tube station.

UCO122: Which didn't work.

DI Wells: That depends on how you look at it.

[PAUSE]

Sometimes we need to take the long view, Daniel. It might not be such a bad thing that they were able to cause so much damage.

UCO122: Sorry?

DI Wells: Think about it. These groups are still seen in some quarters as a bit, well, fluffy. The national news coverage of this event will make clear how important it is that we continue to keep them under surveillance.

UCO122: [LAUGHS] Well, yes, but obviously that wouldn't be a reason for us to allow things to get out of hand.

DI Wells: Of course not. No, I'm just pointing out that it could have some helpful side-effects, from our point of view.

UCO122: I see. That's not quite . . .

[PAUSE]

You know what? Nothing. I have to go. My arm hurts.

DI Wells: Of course. Let's speak again once things have settled down.

II.

1997–9

5. MAINTAINING COVER

5.6 SEXUAL LIAISONS

5.6.1. The thorny issue of romantic entangle-
ments during a tour is the cause of much
soul-searching and concern. In the past emotional
ties to the opposition have happened and caused
all sorts of difficulties, including divorce, decep-
tion and disciplinary charges. While it is not my
place to moralise, one should try to avoid the
opposite sex for as long as possible.

[...]

5.6.3. While you may try to avoid any sexual
encounter there may come a time when your lack
of interest is suspicious. [REDACTED. Gist: 'this
sentence provides advice to UCOs not being in a
sexual relationship with a member of the group'].

These options are fraught with difficulty and you must make your own mind up about how to proceed. If you have no other option but to become involved with a weary, you should try to have fleeting, disastrous relationships with individuals who are not important to your sources of information. One cannot be involved with a weary in a relationship for any period of time without risking serious consequences.

Special Demonstration Squad,
Tradecraft Manual

1.

'**O**KAY, I have a question.'

'Milady?'

'Can I throw this away?' She was holding up a poster in a cheap clip frame. It showed the head of Maggie Thatcher with an axe embedded in it, and a pool of blood beneath the words: CLASS WAR. This was one of the few items Dan had brought with him to Heron Court when he moved in. He didn't have much else: two hold-alls full of clothes, a Sainsbury's carrier bag with a few CDs and tapes (confirming his musical taste to be terrible: not only Bon Jovi, but also Meatloaf – Meatloaf!), and a yellow plastic box. *I'm not really a things person.*

He pulled a sad face. 'You don't like my poster?'

'It's grim.'

'Chuck it,' he said. 'I don't care.'

'But she was still looking at it, looking at him, then again at the poster.

'It's also really *not you*.'

113

'What do you mean?'

'It's so strident. I really can't imagine you seeing this and thinking, *That's the one I want on my wall.*'

He laughed. 'You're reading quite a lot into this. I really didn't think about it that hard. It's just a poster. How about this one?' He held up a giant framed reproduction of a photo that had appeared in the paper last year, the weekend after Shepherd's Bush: Aoife, imperious and resplendent in her Marie Antoinette tent-dress, waving to the raving crowd, with a row of black-and-white police hats at the bottom of the frame. Dan had loved it so much that he sent off to the paper for it (she hadn't dared to ask how much *that* cost).

'Now that,' she said, 'is quality artwork. That one gets pride of place.'

'Here?' He held it up in the centre of the newly decorated living-room wall.

'That's the spot.' She walked to the hall, organising things in her mind. 'I thought we could put Tofu Love Frogs here, and then just have photos all over this wall. And I want one of those corkboards in the kitchen, just for random postcards and notes and stuff.'

It hadn't been a difficult decision, in the end. Dan had been staying at Heron Court pretty much all the time since last summer. There was no point, they had agreed, in her continuing to spend almost her entire play-scheme earnings on rent, while he paid out for his crappy flat above a kebab shop on Manor Road. She'd only visited him there twice: it wasn't exactly an inviting place. Typical bloke, he had put absolutely no effort into making it nice: it had chipped magnolia walls, a mock-leather sofa and a hi-fi, and that was pretty much it. Not a single cosy cushion or lamp, no photos on the walls, just the Class War poster, which she had been secretly hoping he wouldn't bring with him.

She moved it to the charity-shop pile, and stood back to survey their progress. The flat was – finally – beginning to look like a home. The decorating was finished. Unlike any other man in her life, Dan had proved to be very useful in this respect – even if he had vetoed her choice of turquoise and orange, and insisted on painting all the walls and the woodwork white. He had also bought a sofa, a coffee-table, a dreamily comfortable double bed and a stupid big telly (all from John Lewis! The price!).

But although she didn't always share his taste, she had to admit that, now it was nearly finished, she liked it: the place looked spare and clean and kind of grown-up.

Perhaps that was what they were, now – grown-ups. It was odd, how it had happened. It seemed like only yesterday that she'd been living in a tree-house, drinking cider for breakfast. Now, she found herself turning into a fluffy little domesticated hen. Dan was away a couple of days a week for work, as his job took him all over the country. She loved those days, when she could sweep and tidy all the dumbbells and Rizla and bits of screwdriver and other man-junk into Dan's designated cupboard and make the surfaces of their little flat gleam, then put her slippers on and feet up, and feel tidy and complete all by herself.

Her gaze reached the far side of the room, and stopped. There was one thing still bothering her. 'Those shelves you put up are totally on the wonk.' She went over and fiddled with them. 'Are you sure you're in the construction business?'

'Oy!' He threw a cushion at her; she reached over and tickled him, and they fell together laughing onto the brand new butter-soft sofa. Leaning his delicious weight on her, he nodded up at the photo, now hanging on the wall. 'The place looks good, though, eh?'

'It looks great,' she said. 'It looks like ours.'

Not that everything was pink clouds and happily-ever-after. There were times, odd times, when something bothered her. Times when Dan could be cold, or distant. Often she felt him to be there, physically, in the room, but with his attention somewhere else entirely. It was a small thing, nothing to write home about, nothing compared to being nearly strangled or living with a drugs-and-sex-hotline addict. Nobody's perfect, she told herself. There's always going to be something. But . . .

'You know, now and again,' she said, as they lay one evening entangled in the Egyptian cotton sheets on the dreamy new bed, 'I'd like us to fuck with the lights on.'

He had been reaching out to turn off the overhead light, having already completed his rigorous curtain-drawing operation, but now he froze halfway there, dropped his arm. 'Oh.'

She turned his face towards hers, looked straight into his dark eyes with her clear bright blue-green ones. Kissed his lips. 'I want to see you!'

He shook his head free. She was finding out that, at least when sober, he wasn't good with this kind of open-hearted comment, or with any attempt by her to connect with his inner core. It was just a man thing, probably: they had to protect themselves, pretend they were tough. His body told her otherwise, fingers softly playing with her curly hair. She breathed in his minty-earthy smell. He was always so clean, always washed before and after sex, which she was still trying to get used to, having been traditionally on the natural-body tip herself.

Soon after moving in, he had bought her a razor as a present. It was one for ladies, with a pink handle. She had burst out laughing, never having shaved her legs or pits in her life. *Is this a hint, Dan? Are you finding my natural body hair problematic?*

All right, Emmeline Pankhurst. It was on offer, that's all.

She disentangled herself, sat up, hugging her knees to her chest. He rolled onto his back, fixed his eyes on the ceiling, as if there were something more interesting up there than council-issue polystyrene tiles. 'Don't you want to see me?'

He raised a suggestive eyebrow, put on a posh accent. 'I have a vivid imagination.' She didn't laugh. He clocked this. 'It's not that I don't want to. I just – enter into things a little bit more when I can't.'

Tears sprang to her eyes, quite unexpectedly, and she wiped at them with the back of her wrist.

'Hey, don't get upset. It's not personal.' He gave her knee a comforting stroke. 'Blame my parents. It's probably their fault for messing me up.'

She sniffed. 'You say that about them, yeah.' He had sat up, too, now, so they were facing each other. The light hung over his head, as if this were an interrogation. Well, maybe it would be. 'Do you want to invite them over some time? Sidney and Diane? I'd really like to meet them.' He was silent. 'I don't mind what they're like. That's just parents, isn't it? I mean, you've met Mother, she's hardly—'

'Yeah, that's not going to happen.' He rubbed his eyes, which was what he did whenever she tried to talk to him about things that weren't just everyday.

'Really? Never? Why not?'

A flush was gathering at the base of his neck. 'I've told you, Sky. You know what happened, with my dad. The booze has wrecked him. He's not somebody I can have over for a cosy little Sunday lunch.'

She reached out and stroked his hair. 'I'm so sorry.'

'It is what it is.'

For a moment she waited, stroking, wondering whether to leave it there. But she couldn't. 'You've told me all this stuff. The facts: your mum dying, your dad losing it, you running off to Spain, and then prison . . . But I don't know. It all feels a bit distant, when you talk about it. You never say how any of it made you feel.'

He laughed uneasily. 'I'm a bloke, Sky. I don't do feelings.'

'Bollocks.'

'Except for you,' he said quickly. 'I have feelings for you, obviously.' Redness was rising up his neck, into his face, and it occurred to her that she should have initiated this conversation when they were drunk or high on happy-drugs. That might have gone better. But somehow, when they were high, when they were completely in the moment, she didn't care enough to do it.

'I'm sorry, Sky. I don't know what you want me to say.' His face was burning red now. He rubbed his eyes again. He looked so horribly uncomfortable that she didn't know whether to go on, but found herself doing it . . .

'I guess,' she said carefully, 'with the lights-off thing, it feels like another way of, I don't know, keeping me out. Like there is part of you that you want to keep secret.'

He just sat there, squinting slightly in the light. Then he was rubbing his eyes again. He was fumbling on the bedside table for his smoking stuff.

'You look like you'd rather be anywhere else on the earth right now than here with me having this conversation.'

'Bullseye,' he said. 'There you go, see? I'm perfectly well in touch with my feelings.'

'Dan, please,' she said softly, 'this is important.' She leaned forward, took his hands and held them, fixed his eyes with hers and wouldn't let them go. 'I don't mind what it is, the secret thing. I just mind that you're hiding it.'

He turned back to the smoking gear, couldn't find the filters, opened the drawer and rummaged through it. They weren't there. He stopped, looked down. 'Okay. You want to know something about me. Here you go. There is only one person I consider to be family, and that is Adrienne.'

Ah! She was on the scent now, with this new name. A crack was opening, a crack in the wall.

'Who,' she asked very gently, so as not to alarm him, 'is Adrienne?'

'My aunt,' he said. 'She lives in Spain.' He took another deep breath, seemed to gather the will to carry on talking from somewhere. 'When things got really bad with my dad, when I was seventeen, Adrienne was the one who got me out. She's the only one that's looked out for me.'

'And are you in touch, now?'

'Now and then, yeah. When we can. But as she's abroad it's not exactly easy to see her. Anyway.' He covered her hands with one of his, pressed them together, as if he was praying or something. 'That's enough. I don't think the past matters,' he said. 'We're both here now. That's all.'

Having finally gathered together the necessaries, he stood up, heading out onto the balcony for a smoke. He was halfway out of the bedroom door when he paused, standing with his back to her. 'You might just have to accept the darkness thing. To accept me, how I am. Sorry.' He said this without looking around, then left the room, closing the door softly behind him.

2.

WORD had spread about the world-changing group. After Shepherd's Bush, everybody had a view with an exclamation mark next to it.

Terrorists! said some. Delinquents and vandals!

The future of the Left! said others. The return of the radical!

'Nobody gets it, do they?' Rev said despairingly to Skylark, in quiet moments. 'Or, rather, everybody who was there got it, completely. But then afterwards it all gets mangled, translated, interpreted . . .'

The Tuesday-night meeting was no longer ten random misfits in Rev's studio. So many people started turning up that they didn't fit there any more, and Dan had to hire a hall down the road. A lot of the new recruits had glasses and beanies and master's degrees in sociology. *Hegemony*, they would say, or *superstructure* or *appropriation*. Most of them were called Gaz, and even those who weren't might as well have been. They bonded with Jez, who shared their love of long words, and formed a group-within-a-group that we shall call the Jez-Gazzes.

The Jez-Gazzes were Very Serious about Changing the World.

'Yes, yes, it's all very well, this rave stuff, all these parties and repetitive beats,' they said, once they had settled in, found the biscuits and made themselves comfortable, 'but what about the global poor? What about trade negotiations and agricultural subsidies and the defence of human rights in the developing world?'

'Hmm, well, yeah,' said the Old Guard, the hippies, the lunch-outs, the squat-bunnies. 'Our aim was more to shoot an arrow of hope . . .'

'A whattie of what?' cried the Jez-Gazzes. 'Don't you see? It's no use getting the masses off their rocks on pills and dancing! It's no good just giving them free dahl and drumming circles! What good are fabric sunflowers to anybody, if they don't understand the nature of their exploitation? We need to educate! They need to be taught about clauses and sub-clauses, committees and working groups! About turtle-friendly fishing practices and GM crops and also something called neo-liberalism!'

'Oh,' said the Old Guard. 'So what's that last one, then?'

And at that the Jez-Gazzes would fall silent, as they

didn't actually know, or they did, they absolutely understood what neo-liberalism was, of course they did. It was just hard to explain.

Arrows of hope and suchlike were thus relegated to the back-burner, while the Jez-Gazzes banged up the heat on the politics. They linked up with the dockers' unions, transport unions, formulated views on transport policy. And with Labour having won the election only months ago with a landslide, there was a whole new establishment to negotiate.

'Obviously we don't buy any of the Blairist bullshit,' announced Rev, loftily, seated one Tuesday on the Throne, which was now a plastic chair in a bigger-than-before circle in the hired hall. 'The man has a soul of darkness. Just look at his teeth – far too white. Those teeth tell me everything I need to know.'

'Don't laugh,' stated Bendy Aoife, 'but he's quite hot, for a politician. And aren't they better than the Tories, though?'

'Politricksters,' intoned Mouse, 'ain't got no souls.'

This discussion was followed in baffled silence by the Jez-Gazzes.

'Dan,' said Rev, 'are you getting all this down?'

He was. Dan always took the minutes. He filed them away in a special cabinet he had put in the spare room at Heron Court. Dan had proved himself to be very efficient, dedicated. He did the accounts, and kept not one but several databases of names, addresses and telephone numbers, arranged into trees, so hundreds of supporters could be called upon within minutes in an emergency. He booked the room for Tuesdays, and made sure they paid for it. The Jez-Gazzes had warmed to Dan. Dan, they agreed, was a responsible and mature individual, somebody who could Get Things Done. If it weren't for him, the whole thing would degenerate into a Complete and Utter Hippie Shambles.

'Thank God for Dan,' Rev said to Skylark once. 'He volunteers for all the most boring jobs, the ones none of us can be arsed to do. Doesn't he mind?' But he didn't. He was practical, and she admired him for that. Dan, who was level, Dan, who kept everything under control. Practical Dan.

3.

'HONESTLY, chicken,' said Suze, 'I don't know where you find the energy. Changing the entire world, it's not exactly a small task, is it? You've chosen quite a hobby.'

Skylark yawned and put another sugar in her coffee. It was true. With all the meetings, the admin, the endless to-and-fro of different opinions and ideologies, world-changing had been a blast, a dream sprinkled with fairy dust and starlight, back when it was all fun and games and she didn't have a care in the world. Now, she had a flat to keep up and a nine-to-five job and a boyfriend. She didn't want to listen to the Gazzes droning on about hegemony after eight hours at The Crew. She didn't want to spend every weekend with Dan driving people and tat around between warehouses.

'It's not like it used to be,' she had found herself griping to Rev. 'It feels like work, now.'

'Tell me about it,' he agreed. 'Try making art when a

thousand people are trying to lecture you about transport policy.'

Suze was faffing around at her desk, shuffling through mountains of papers. 'I'm just trying to find the latest letter from Tessa,' she said, with a significant twinkle. 'Our new Labour government's minister for families.'

'I take it there have been developments?'

'As a matter of fact, yes, I saw her last week,' said Suze, adjusting her glasses. Her giant tasselled earrings jingled dramatically. 'Just between you and me' – she leaned forward, dropping her voice to a stage whisper – 'she's promised us money to build a whole new centre. Computers, therapeutic swimming pool, adventure playground. It will be State of the Art.'

'Suze! How did you talk her into that?'

Suze frowned momentarily, then flashed her gaudy, reckless smile. 'I never talk anyone into anything,' she said. 'I just showed her what we do with these children, the difference we can make. She's a good woman, and she can see it with her own eyes. We change lives here every single day. Once you open somebody's heart, good things just flow naturally. Now, off you go, Little Miss Revolution. Get back to work.'

You had to hand it to Suze, Skylark thought, as she left the office, clutching her coffee. The woman knew how to get what she wanted. When it came to it, who was really going to make things better? The world-changing rabble, with all their big ideas, or Suze, with her negotiating and her networking, her willingness to locate and pull the levers of power? The Crew could be tiring and menial and overwhelming, but at least it was real. No ideological squabbles involved in arse-wiping or medication-administering, just tasks to be done.

Jase was sitting in his usual place on the crash mats. Over the last year puberty had set in. She couldn't lift him any more, and he had stray wisps of dark hair on his chin. He'd also developed an annoying new habit of lunging for her chest whenever he got the chance.

'All right, big man?' He looked at her, with his faraway eyes. 'Wanna do some rocking?'

He took her hands, held them. She sat still for a moment, as a thought arrived in her mind. Then, as they started to rock, it settled gradually down, cemented itself, revealed its clear and obvious nature. 'You know what, Jase?' she said. 'I think I'm getting to the end of the road with this world-changing lark. I've seen too many burnouts. I don't want to be one.'

'Bah,' said Jase. 'Dah.'

'Yeah. You know? Maybe soon it will be time to move on.'

* * *

She'd been looking forward to having the flat to herself that evening. Dan had left for his usual working days away, some building site up near— Where was it, this time? Manchester? Carlisle? She hadn't really been listening.

She planned to spend a while writing a letter to Rev. *I'm sorry I can't do it any more, Revvy*, it would say. *This thing we've made together has been the most important part of my life. I've given it everything, and it's given me everything back. But time moves on, and I'm building the next phase of my life. I still want to change the world. But I'll do it in a different way. A way that is more about people, and about love, and less about grand ideas. But I'll still be here, willing you all on. The flaming arrow, for ever!*

These words were scrolling through her head as she turned the key in the lock and pushed open the door. And then she stopped, her heart giving a twitchy little bump, because somebody was inside.

She pushed the door open, a little wider. No, she wasn't imagining it. It was a noise, a somehow aggressive and unpleasant scratching, coming from the kitchen.

Scritchscritchscritch.

Scritchscritchscritch.

'Dan?' she called. 'Dan, is that you?'

There was no reply.

Scritchscritchscritch.

Scritchscritchscritch.

She stood there, hesitating on the doorstep. Nobody else had her keys. Who could possibly be in her flat? Although Heron Court was on the rough and ready side, she had never worried much about breaking and entering. Until Dan had started buying all this posh stuff, she'd never had anything to steal.

Scritchscritchscritch.

Scritchscritchscritch.

It sounded almost like an animal. Had a dog or a cat got in somehow, through a window maybe? She turned her keys in her hand, so that one was jabbing outwards. As she did this, she knew she was being ridiculous, that keys would be no use at all if the scratching thing inside was a criminal, a rapist, or a mad dog. She crept down the hall.

Scritchscritchscritch.

Scritchscritchscritch.

She craned her neck around the door to see into the kitchen, and an adrenalin jolt shot through her. A man was standing there, with his back to her. A soldier. There was only a fucking soldier in her kitchen. He was leaning over something, his arm was moving.

Scritchscritchscritch.

'Hey!' she shouted wildly, recklessly, brandishing her useless ridiculous keys. 'Get out!'

The soldier turned. It was Dan. It was Dan, holding his old pair of walking boots, and a brush. 'What the—'

'Oh, my God,' she breathed, hand on her hammering chest. 'I just – I didn't recognise you.'

'What?' A flicker of something passed over his face, before he grinned. 'Crazy lady,' he said, and then, when he saw her expression, 'Were you really scared?'

He'd been polishing his boots. *Scritchscritch*. Dan. It was only Dan. The man she lived with, the man she saw every day. She leaned against the doorframe, breathing, closed her eyes. Things were spinning.

'Work got cancelled,' he was saying. 'They've had a problem with the planning and it's all held up. Hey . . .' He touched her arm and she opened her eyes, shook the alarm and confusion out of her head. She must be really tired, or losing it completely. He drew her into a hug. 'Sorry. I didn't mean to freak you out.'

'That was so strange,' she said, over the noise of an alarm still going off somewhere in her head. 'For a moment, you didn't look like you, at all.'

4.

THE Tories were gone and it was free party season.

Skylark found herself with Bendy Aoife one Saturday afternoon on Walthamstow Marshes, in the May sunshine, dancing to reggae as estate kids in shell suits smoked brazen spliffs, and confused butterflies trembled in the vibrations. Someone-who-was-friends-with-Rev had set up a sound-system, because why not? A bunch of police hung anxiously around on the margins, trying to look like they had it all under control.

Dan wasn't there: he'd been away working, but she'd given him directions. He'd said he'd come along when he got back. Unfortunately the afternoon was proceeding a little more quickly than she had anticipated. Someone-else-who-was-friends-with-Rev had given her and Aoife a bottle of rum and ginger, then spotted yet another someone, whom they hadn't seen since Oldbury, and wandered off. So naturally the two of them had finished the whole thing. The afternoon woozed; her head was floating somewhere above her body.

It was five past three.

'Let's dance at the front,' said Aoife, and they stood right next to the speakers, bass rattling their bones. Someone passed her a joint and she smoked it while she danced, long green tendrils of skunk reaching down into her veins, so that the music merged with her movements in a way that felt good, felt right, even though she had a definite sense of being too messed up, too early in the afternoon.

It was at that point that she felt a poke her between her shoulder blades.

'Skylark,' said Mikey, or mouthed it really: the music was too loud to hear. He'd cut his hair: it was short and tousled now instead of long and matted, which made him look younger and more together, and he was wearing the black-and-red-striped Dennis-the-Menace jumper he'd always had on back in the camps. She used to love him in that jumper: it made him look like Kurt Cobain. In the haze of cider and skunk, he appeared to her once again, like an apparition, as the old charismatic Mikey, the man who had guided her into the world-changing life, and who had been the original dreamer of their world-changing dream.

* * *

Because – although now this was passed over and forgotten by the other world-changers – it was Mikey, in fact, who first had the vision. This was in Harfield Road, the squat where they had lived before Heron Court.

Harfield Road was in Leyton, east London, and had been scheduled for demolition. The proposal from the powers-that-be was to knock down this humble little community, with its terraced houses and leafy trees, to bulldoze this quiet and unassuming human habitation, and build instead a monstrous stretch of tarmac called the M11 link road, a motorway making an incursion right into the inner city.

Once the powers-that-be had moved the residents out, she and Mikey and Rev and Aoife and the whole damn lot of them had moved in, a collection of former tree-dwellers, hippies, Travellers and shit-stirrers from all over the country. They had squatted that entire street, and made themselves a miniature little betterworld right there. Painted the houses with strings of daisies, giant chessboards on the tarmac, opened a jazz café in what had been someone's front room. Partied all night, slept all day, tangled together, like nesting animals, in cargo nets strung from the rooftops.

Because there were no cars, they dragged abandoned furniture out into the street, made their sitting room on the road itself. They ate together out there, around an evening fire. There was a play area for kids, with swings and a climbing frame.

One evening, sitting around the fire with beers under the purpling darkening evening sky, Mikey had called for silence. His charisma back then was such that silence immediately fell. He swaggered to his feet, the outsized jumper falling off his shoulder, eyes sparkling with mischief and reflected firelight. 'My dear friends,' he said, 'a moment of quiet, if you please, for the best sound in the world.'

Ceremonially, he held up a bottle of Jameson, and with his thumb worked the stopper. It came loose with a soft, satisfying *ploomp*. He held it aloft, then tipped the bottle to his lips and swigged. There was a muted cheer from his audience and Mikey bowed, before sending the bottle to his right. But he stayed standing up, rock-star-style.

'Comrades, brothers, sisters, old muckers, I've been imagining,' he said, 'and I wonder if you can imagine, too. This lifestyle we have here – you will agree – is a glorious thing. It may not be glittering, it may not be caviar and limousines and Buckingham fucking Palace.'

('Off With Their Heads,' heckled someone, to another cheer.)

'But we, my friends,' Mikey went on, 'have a freedom

that even our esteemed Royals can only dream of. And why should we keep this marvellous secret to ourselves? Don't we have, in fact, a responsibility to share it with the world? For the sake of a betterworld? After all –' he staggered slightly, regained his balance '– if we stopped the traffic, everyone could live like this. And I think ordinary people deserve to know that.'

It was Mikey, who'd dreamed it all up. Mikey, who'd never asked for, cared about or got any credit.

'Skylark,' he said, again. 'It's been a while.'

'You cut your hair,' she said.

'Yeah, well,' he sized her up, 'you look just the same.'

He beckoned her to come after him, further away from the music. Aoife shot her an exasperated warning look, but she ignored it. Fuck it, really. She was caned, and it was summer.

'Where've you been, then?' she asked, once they'd sat down in the grass, Mikey's jumbo bottle of Thatchers propped between them.

'Ireland,' he said. 'Giant's Mountain.'

They'd been to this encampment, a Traveller site populated by blow-ins from England, a couple of years before. They sat for a while talking about old-time stuff: who'd come and who'd gone. A couple called Dave and Ali, old friends of Moll, whose wagon burned to the ground on his first night back, he told her. A spark from the wood-burner must have landed on the canvas top, and by morning there was nothing left of it but a pile of ash.

'They were so gone they nearly slept through the whole thing. Didn't wake up until their eyebrows were on fire.'

'Hey.' Their laughter stopped as someone stood in front of them. She squinted into the sun, knew who it was instantly. Nobody else had a silhouette that broad.

'Oh, hey, Dan,' she said. 'You're here. This is Mikey.'

Dan put out his hand and Mikey shook it. Then he turned and frowned at her. 'You all right, Sky?'

'Yeah.' She was only slurring a little. 'I'm good, having fun. Summer, and all that. It's nice. Have a drink!'

He put a hand on her shoulder. 'Why don't we go for a little walk?'

Mikey had never met Dan before, but had maybe heard about him from the rumours and gossip that spread rapidly around the scene. 'Sorry, mate, is there a problem?' he asked.

Dan looked at him, up and down, and his expression made crystal clear what species of worm or dung-beetle he thought he was looking at. Skylark had never seen that look before: pure venom. He shook a finger. 'I know who you are.'

Mikey hopped to his feet. 'Sky? What's he on about?'

Dan took her by the arm, tried to drag her up to a standing position. 'Come with me,' he said. 'You're drunk.' But she shook him off.

'Who is this joker?' asked Mikey. 'Look at him, he's like Popeye, with those muscles. He don't belong here. Hey, come here, wait, let me look at you . . .' He leaned in close to Dan, narrowing his eyes. 'Are you the law or something?'

Thunder came over Dan's expression. He moved towards Mikey, as Skylark stumbled to her feet, clumsily tried to hold him back. Next to Dan, Mikey might as well have been made from bones and fluff – he didn't have a hope. Dan grabbed him by the jumper's baggy collar. 'Hear me? I know *exactly* who you are,' he said, again.

Mikey swung, but Dan caught his hand and, in a smooth motion, flipped him onto his back. He knelt down, face inches away from Mikey's, forearm resting heavily on his neck. Leaned in close. 'Stay away from her. Understand?'

Then he was up and off. As Mikey caught his breath, swearing and rubbing his neck, Dan took Skylark's hand firmly and marched her away.

DI Wells: You'll be relieved to know that, despite the election result, I think we're going to be okay.

UCO122: Our new leaders aren't getting the knife out?

DI Wells: It appears not. The current leadership are very keen on public order. They need to reassure all those former Tory voters. And nothing alarms Mondeo Man more than a bunch of wearies ranting on about bringing down capitalism. They're going to stick to the Tory spending on criminal justice – for now.

UCO122: That's good news.

DI Wells: Absolutely. In fact, word is, there'll be more investment coming our way over the next couple of years. And they're talking about new legislation, an update of the Criminal Justice Act. They're going to reclassify some of our wearies as 'domestic extremists', which will make things much easier for us.

UCO122: Ah.

DI Wells: Reservations?

UCO122: Well, I can see the advantages to it. But between you and me, Martin, extremists? Really?

DI Wells: [LAUGHS] I could point to a drilled-up motorway.

UCO122: There was that.

DI Wells: And, besides, you may not want to make that point too loudly, if you don't want to be out of a job.

UCO122: [LAUGHS] Right.

DI Wells: Oh, yes, and one other thing: they're talking about rolling out our activities beyond London. Setting up a new unit to cover the rest of the country. And any new recruits will need some training, so . . . I'll keep you posted.

UCO122: Right. Because obviously many serious threats to the status quo are brewing in, say, Swansea and Tunbridge Wells.

DI Wells: You may mock, Daniel, but these are our orders. It's not for us to wonder why.

[PAUSE]

Okay. So what are they up to, then?

UCO122: Well. An interesting opportunity has arisen.

DI Wells: Tell me.

UCO122: It's something of a strange tale, Martin. On Tuesday after the meeting I went for a drink with Rev – Rupert Delamere.

DI Wells: Lord Delamere, to us plebs.

UCO122: Wait for it, though. He tells me that last week he got called in for a meeting at the Cuban embassy.

DI Wells: You what?

UCO122: I'm not kidding. To be fair, he was pretty surprised himself. He's hardly a guerrilla fighter, after all – I don't think the guy would survive ten minutes out in the Sierra, in his frock coats and pointy shoes. Anyway, he turns up, and gets ushered straight in to see the ambassador. Who issues him with an invitation. I've got it here, have a look.

[PAUSE]

DI Wells: *'Encuentro por la humanidad y contra el imperialismo.'*

UCO122: Nice accent.

DI Wells: I did Spanish O level. Let me see – what is it? *Encuentro*? Is that a meeting?

UCO122: It's basically an international get-together of revolutionary groups – from the Zapatistas to the Naxalites to the Tamil Tigers to the Shining Path. Happening in Spain this summer. Organised by some Mexican insurgents, backed by Castro.

DI Wells: Is this real?

UCO122: It seems so. The crazy thing is, that Rev – Rupert – has decided not to go. Says he has no interest in being part of some pre-existing political agenda, be it left or right.

DI Wells: Could be a nice holiday, though.

UCO122: He didn't seem to see it that way.

[PAUSE]

I, on the other hand . . .

DI Wells: It could be very interesting.

UCO122: Exactly. I mean, if I went, I'd get access to pretty much every left-wing revolutionary group in the world. Obviously they're planning to start working together, on something international, so . . .

DI Wells: It would be complicated, moving you out of the country. I'd have to clear it with the Spanish police.

UCO122: There are some new wearies in the movement, joined since the M41. They're much keener to take things in a more explicitly political direction. I think they might well be quite pleased if I offered to go, as a kind of representative in the absence of Rev.

DI Wells: Right. Leave this with me. I'll need to get someone out there with you, for back-up . . . maybe Nigel. He can keep an eye, make sure you're not having too much fun. I guess you'll try to squeeze in a little family visit while you're out there?

[PAUSE]

UCO122: I don't know, Martin. Between you and me, Adrienne is not too keen on visits, just now.

DI Wells: Oh? What – is that still going on? I thought you'd patched it up.

UCO122: Not really, no. I mean, we came to an understanding.

DI Wells: An understanding that you wouldn't visit?

UCO122: That was pretty much it, yes.

DI Wells: Daniel, you should have told me.

UCO122: Why? It's my personal business. Nothing to do with work.

DI Wells: As you know, undercover officers are always married men. It's a requirement of the job. Otherwise things can get very confusing.

UCO122: Yes, but—

DI Wells: No buts, Daniel. You should have told me.

[PAUSE]

Anyway. Let's get this little Spanish trip sorted out. That is mission critical. This is very good work, Daniel. Bob will be delighted.

5.

A SHOOTING star streaked across the sky, like a firework, brighter and closer than any she had seen before. A murmur rippled around the crowd, and she held on more tightly to Dan's arm. A man walking next to her tapped her on the shoulder, pointing upwards. '*Estrella fugaz,*' he said. '*Buena suerte.*' He was small and dark, with a compact, hardy body. He had a guitar slung over his shoulder, and his eyes shone despite the dark.

'Good luck,' she said, because she understood that much. 'Yes.'

'*Pachamama nos bendiga.*' He put his hands together in a prayer position, and she did the same, hating the fact that she couldn't speak the language. Dan looked confused, and a short conversation followed in his halting Spanish.

'I think he's saying Pachamama is, like, an earth goddess.' The little slanty smile flickered on his lips. 'And that she blesses us.'

'That's lovely,' said Sky, smiling at the small dark man. Then to Dan, 'Don't you think, Danny?'

'Oh, please,' he muttered. 'I'm not that far gone.'

She nudged him. 'Who are you kidding? You're here, aren't you? You're officially the furthest of the gonnest.'

The Encuentro was taking place in the Andalucían countryside, in a farm that had been squatted by Spanish land-rights activists. During the day, the sun beat down on the olive groves, and the parched brown hills above. The farm was a huddle of whitewashed barns and buildings around a small stone courtyard. On one side was the kitchen barn, where a group of hearty *camaradas* served up fried eggs, sangria, and lashings of earthy stew – like Moll, with added paprika. Across the courtyard was the main barn, where the *seminários* took place. There had been *seminários* every morning and afternoon, on such topics as 'building a feminist revolution' and 'constructing global networks of communication to support the struggle'.

Sky had struggled to remain awake through the *seminários*, which were in Spanish, with trilingual translation, and revolved around certain key phrases that cropped up with numbing frequency: *neo-liberalismo, capitalistas, trabajadores, la lucha*. It didn't help that every evening had been spent drinking throat-stripping *aguardiente* around a bonfire and singing revolutionary songs from around the world with ever more ear-jangling tunelessness.

But now it was the last night, and the whole Encuentro – nearly five hundred people, from all over the world – was processing down an unlit road into the depths of the Spanish countryside, towards an event described as 'closing address from the *comandante* and celebration of international revolutionary consciousness'. There was something truly magical about the moment: all these people, of all colours, nationalities and languages, walking through the heavy hot night, following the light of flaming torches. She took a deep breath of warm Spanish air, which tasted so foreign, heavy with red clay and sticky sap. It entered her lungs, and her bloodstream absorbed it, so it enveloped her inside and out.

Dan's decision to come to the Encuentro had not been without controversy. It had signalled a new dynamic in the world-changing group; until that point, Dan's role had been as the wingman, the transport guy, the dogsbody, the driver. He was the person you could rely on to do as he was told. But Dan, it seemed, was no longer simply content to drive his van from A to B. He had challenged Rev.

'I mean, our agenda is anti-capitalist, right?' he had said, during a tense Tuesday meeting.

'Of course,' said Rev. He looked shocked, hurt: this was like getting told off by his kid brother.

'And these international revolutionaries are also anti-capitalists, right? In that they're fighting the dominant economic model in their own countries?'

'Right.'

'So – you're on the same side. We need somebody to go – to hear what they have to say.'

This was catnip to the Jez-Gazzes. It was just what they were looking for, with the zeal of the newly converted: a puffed-up sense of their own Great Historical Importance.

'Yes!' they cried. 'Think bigger! Create an international network! Take on the forces of global capitalism! Bring down the system!'

'But,' said the Old Guard, 'the arrow of hope. The party as a portal.'

'Old news!' said Dan, Jez and the Jez-Gazzes. 'Never mind organising frivolous drug-taking parties with your lunched-out friends! While you lot are tinkering around with repetitive beats, we're going to link up with our international revolutionary brethren and Change the World!'

So eventually, inexorably, the red-and-gold invitation

had come to Dan. And then, later, he had suggested that Skylark should come, too.

'Just for the holiday. Have you ever been to Spain?' She shook her head. 'Andalucía is my favourite place in the world.'

'Hey,' she said. 'That's where Adrienne is, right? Can we go and see her?'

'Er, maybe.' He rubbed his eyes. 'Yeah. I'll ask.'

'Where does she live, exactly?'

'Jerez,' he said. 'A little town called Jerez de la Frontera.'

Now she had the bit between her teeth. This was, finally, a chance to meet the famous Adrienne, the only person Dan considered to be family. She went and bought a guidebook, looked up Jerez on the map.

'Look,' she pointed out, 'it's really close to where we're going. Jerez is just down the coast from the Encuentro, about twenty miles. So we should definitely go and see Adrienne while we're there.'

He looked at her a bit wearily. 'Well, yeah, we can try, I guess, you never know.'

'Will you ask her? Definitely?'

'Sure. Of course.'

But then just before they left, she asked him again about this plan, and he said, 'Oh, yeah, sorry. I meant to tell you. We can't, Adrienne's away. She's not in the country.'

She was crestfallen. So disappointed that she almost couldn't bear to go. But by then he had made all the preparations: bought a mattress and a cooker to put in the van, and tickets for the ferry from Plymouth to Santander. Before they left, he was careful to put up nice thick curtains around the van's windows, so they could make it dark.

'*Cuídate, chica*,' said the man with the guitar. '*Ahí esta el río*.' He was gesturing to where the dry earth gave way to a wide, shallow river. In front of them people were wading; it seemed the whole procession was going to have to cross. Skylark stopped for a moment to take off her sandals. Sank her feet into the cool water, and kept her eyes ahead.

They had been down to the river before, yesterday, after sitting at lunch in the barn with a group of German squatters, who never smiled but spoke very good English.

'It's too hot for politics, no?' said one of the women, with a green mohawk and a pierced lip. She lowered her voice. 'I heard there's a river. About ten minutes' walk away.'

That was all Skylark had needed to hear. There was no question of her returning in the afternoon for the *seminário* on 'the world system for capitalism and its inevitable downfall', no matter how much the international revolutionary avant-garde needed her.

Turned out that most of the international revolutionary avant-garde felt the same. Dan didn't take much persuading to sneak off after lunch, bunking the downfall of capitalism, and following the mohawk woman down through the olive groves. Skylark took his hand as they walked between the trees, scorching sun on their shoulders, pebbles in their sandals. He had been in a silent mood for a lot of the holiday, tense and absent. She was relieved when he gave her hand a squeeze now.

'You all right?' she asked.

'Yeah,' he said, with a fleeting smile. 'Just tired. I didn't sleep too well.'

Down at the bottom of the grove there was a shaded clearing, rough rocky grass and, deep in a hollow, a lazy green river overhung with wide-leafed trees. All around, their fellow delegates were sitting drinking, talking, dozing.

'Geronimooooooo!' A tattooed Spanish anarchist, with a dark mullet and ragged denim shorts, took a run up and threw himself off the highest rock, crashing into the water below. There was a scattered round of applause as he surfaced. Others had picked their way more carefully down the rocks, and were swimming in the river or trailing their toes.

Looking around this small group of delegates was like making an anthropological study of humanity in all its forms. In this small shady spot, Polish trade unionists with sunburned chests were sharing beers with Brazilian farmers in smocks and straw hats. A group of indigenous Colombians were sitting very upright on a rock, dressed in white robes with pointed white caps, watching as a couple of punks splashed each other in the water.

Language was broken English mixed with Spanish and charades, but they clearly gravitated towards a natural consensus on the important things, like cool water and beer.

'Hard work, this,' said Dan, lying down on the rough grass, looking more relaxed than he had since they arrived. He pulled a Corona from his pocket and cracked the top off with his teeth.

'Look.' She gestured around them at the people, the trees, the cool water. 'This is what it's all about.'

'Although obviously,' he said, 'we both very much regret missing out on three hours' discussing capitalism's downfall.'

She bent down to kiss him. 'If anyone asks, just blame me.'

One of the punks had got out of the water, and clambered up to sit next to the Colombians on the rock. One took off his little white pointy hat, and perched it on top of the damp spikes of the punk's hair. After some charades, he turned around so the punk could try shaping his long dark mane into a spike.

Skylark stripped off her sarong and vest top, down to the bikini underneath. Dan watched as she climbed up and stood on top of the rock, looking down for a moment at the drop, and the green water beneath.

'*Hágale, pues*,' said the Colombian, in his nasal voice. He was looking a bit bedraggled, after the punk's hair-dressing efforts. '*No tengas miedo.*'

'He says, don't be afraid,' Dan translated, with a grin.

She took a deep breath, and stepped forward into space. For a moment she was weightless, and then, with a nerve-numbing smash, the water caught her in its cool embrace.

A song rippled down the procession that she recognised from the night before, around the fire. It was undoubtedly rousing, but went on for ages in lusty Spanish before getting to the chorus. She and Dan both managed to join in for the two lines they knew:

A las barricadas, a las barricadas!
Por el triunfo de la confederación!

'*Muy bien*,' said the man with the guitar, and then something else in Spanish.

'He says that this is a very important song,' Dan said. 'From the Spanish Civil War.'

'It's very vigorous,' she said approvingly, and then asked the man, 'Are you a singer?'

'A seenger, a traveller, a member of the yuman race,' said the man, in strongly accented English. 'Manu,' he said, holding out his hand.

'Sky,' she said, shaking it, 'and this is Dan.'

The men nodded at one another.

'Sky, Cielo.' Manu looked at her with his bright black eyes. '*Qué bonita.*' He was inquisitive, alert. He wore a stripy red-and-white shirt, and a head wrap with a red tassel. '*Y Dan. Placer conocerles.*'

The crowd was slowing down. There was a small, narrow bridge, and on the other side a sloping field, with a fire burning bright at the bottom. Flames were dotted around the crowd, making the whole scene look like some kind of medieval battle. The throng surged and settled, everyone looking down at the fire, waiting . . .

'What happens now?' she whispered, but Dan didn't know. The crowd was still, expectant. A dark figure appeared by the fire holding a trumpet, placed it to his lips and, with a dramatic motion, arched backwards. A long, quavering note rang out, and a second blast brought the crowd to silence. A second man moved out of the shadows to stand in the light of the flames. A short, stocky figure in military clothes. He wore a cap, and beneath it a black balaclava. Held a pipe in his hand, and took an ostentatious puff, blowing a smoke ring out into the night sky.

'*El comandante.*' A whisper went around the crowd in many different languages. '*El jefe. Dowodka. Der Kommandant.*'

'Who is it?' she whispered.

'It's him,' Dan said. 'The Mexican.'

Then once again the expectant hush.

'*Hermanos y hermanas,*' said the *comandante*, waving his pipe, '*A nombre de los hombres, mujeres, niños y ancianos del ejército de liberación nacional, les damos la bienvenida a la realidad.*'

She looked quizzically at Dan, who whispered a translation. Brothers and sisters, welcome to reality. Manu took over, carried on translating, as the *comandante* spoke.

We meet here because we have a shared dream: a world free from the destructive forces of capitalism and exploitation. We live in the age of an empire that does not call itself an empire: the United States of America is our ruling imperial power. This empire is governed and regulated by three organisations: the International Monetary Fund, the World Bank and the World Trade Organisation. These shadowy institutions govern our world, and they are completely unaccountable. They make their decisions behind closed doors, and they act in the interests of the United States of America. If we are to challenge this empire, these organisations must be our first target.

The *comandante*'s speech continued for a long time. It featured many repetitions of the usual words: *trabajadores, la lucha*, and the ever-popular *neo-liberalismo* . . . 'He's launching a new organisation,' Dan told her. 'It's called the Anti-Imperialist Global Network. An international campaign against the World Trade Organisation.'

'*Al nuevo mundo!*' cried the *comandante*.

'*Al nuevo mundo!*' repeated the crowd.

'*Hasta la victoria!*'

'*Hasta la victoria!*'

'*Que viva la revolución!*'

After the *comandante* had disappeared back into the dark, a hundred bottles of *aguardiente* were pulled out of a hundred pockets, and another fire was lit. The crowd gravitated towards the flames, laughing, talking, singing, drinking.

'I don't know,' she said wearily. 'I might be all fiesta'd out.'

'*Ven,*' said Manu, 'come with me. I know a good quiet place for us to go.'

He led them up the hill, away from the fires. They followed him along a narrow track into the forest above.

Trees closed around them, in a dark jungly embrace. Things moved in thickets, frogs croaked and birds cooed low. She could only just see Manu in front of her, skipping ahead, sprightly as a goat. Just as she was getting tired, the forest opened up. There was a gap in the trees, and a low wooden bench. They could see the field of people spread out below, the bright points of the fires. The moon had risen, and hung above the scene, bright and butter-yellow.

They sat together on the bench in the close Spanish night. She felt that Manu had been waiting for them, the bench had been waiting for them, everyone in the field below had been waiting for them. It was a moment that couldn't have happened in any other way.

Manu took out his guitar and started to pluck a lilting, jaunty tune, like a Gypsy fairground organ. The simple melody fitted the moment so exactly: it was the sound of now, and now, and now.

'Do you remember,' she whispered to Dan, as he played, 'what I told you that time in the boat?'

'Kind of . . . some hippie nonsense.'

'Those moments when you know "The universe wants this."'

'Oh, yes,' he said. 'I remember now.'

'This is one of them. Don't you think?'

He didn't look at her, but drew her closer. 'Perhaps.'

'Hey, love-birds,' said Manu. 'I have a song for you, Cielo.'

They sat there together as the stars shone and the comandante's fires burned and Manu sang a song that was comic and urgent and plaintive all at once.

> Por el cielo camina mi pueblo, por el cielo
> camina la raza
> Por el cielo camina mi pueblo, Pachamama
> te invito a bailar . . .

DI Wells: That's quite a tan.

UCO122: It wasn't a holiday, believe me.

DI Wells: I'm told you got some relaxation in.

UCO122: Who told you that? Bloody Nigel, holed up in a nice hotel down the road while I slept in a cow barn?

DI Wells: He said you managed two briefings. Just about.

UCO122: It was impossible to get away. There were people around me, day and night. If anyone had seen me sneaking out down to the local Holiday Inn, or whatever they're called over there . . .

DI Wells: Yes. I also heard a bit about the people you were surrounded by.

UCO122: Bloody Nigel.

DI Wells: He told me that you had some . . . company.

UCO122: Sorry?

DI Wells: The girl. That she was with you in the van. He saw you together, when you first arrived.

[PAUSE]

UCO122: Oh, Sky. Yeah. She came along.

DI Wells: We did not authorise you to take a weary with you, Daniel. It was hard enough getting permission for you to go abroad. I spent weeks on all the paperwork, getting permission for you to travel. If Bob found out . . .

UCO122: I couldn't go on my own. It wouldn't have been consistent with Daniel Greene's—

DI Wells: You know the law, Daniel. You did not have authorisation. You took the girl along without permission. How am I supposed to explain—

UCO122: Have you seen the intelligence?

[PAUSE]

DI Wells: I am not for a moment disputing that you gained some very helpful and important information on this trip.

UCO122: I got you names and contact details for pretty much every single left-wing revolutionary group in the world. Phone numbers of several prominent Spanish anarchists, including some extremely gnarly types. Plus details about their discussions, their plans for a global day of action next year – if the Spanish police give you any grief, remind them about that.

DI Wells: [SIGHS] I'm not disputing that you did well, Daniel. Very well.

UCO122: And, anyway, do you have any idea how stressful it is? To be in a foreign country, immersed in this stuff, twenty-four seven? I lay awake every night, terrified I'd talk in my sleep, that I'd say something – that one of them would suss me and come at me when I didn't have any cover or back-up—

DI Wells: I understand the challenges. I've been there myself, remember.

UCO122: I couldn't have done it without Sky. A man on his own always invites suspicion. A man with a woman – now that's a whole different story. She was essential.

[PAUSE]

DI Wells: Daniel. Listen to me. This is regarding also our conversation last time. Certain connections can be helpful – essential – in establishing and maintaining your legend. But it would be entirely counter-productive for a connection formed in the field to develop into –

UCO122: [LAUGHS]

DI Wells: – anything else.

[PAUSE]

Do you understand, Daniel?

UCO122: Thank you, Martin. I will bear your advice in
mind.

6.

'WHAT do you think of the name Ché?'
'Just – in general?'

'For a kid.'

This was the first time she had brought up the children-thing, and she did it lightly, as if she was just messing about, but in fact it had definitely been On Her Mind. Partly, or in fact largely, because he did not see the benefit of birth control. 'Oh, come on, Sky, I can't wear those things,' he would say, if she fumbled for the packet in the dark. 'Who wants to fuck a plastic bag?'

He knew she wasn't on the Pill, because she'd tried it at the beginning but it had made her feel sick and really venomously angry, like she would punch and spit on anyone who ever tried to have sex with her again. Which, she supposed, was contraception of a sort. But when she'd told him, months ago now, after they'd got back from Spain, that she was going to stop taking it, he said, 'That's okay, I'll just pull out.'

She raised an eyebrow.

He put on an American-conman accent. 'Hey, trust me, lady, I'm a professional.'

At first, he'd been pretty good about it. But then mistakes were made, more and more. It happened once, and she took the morning-after pill. Then again, just after her period, so she hadn't bothered. But gradually he stopped even saying the whoops-sorry-it-just-felt-too-good thing, and the mistakes came to pass without any comment or discussion, to the point that she thought, although they hadn't actually discussed it, it was obvious . . .

'Ugh. Too political.' He popped a strawberry into his mouth and chewed it. 'Bowie is a good name.'

'Oh, please, Granddad. Not my generation.'

'So, what's a rave-generation name?'

'Um. Ebenezer?'

'That would be child abuse.' There was a pause. 'What – Sky, are you . . .'

'No,' she said hurriedly. 'I'm not. Just thinking about it, you know. For the future.'

It was a Saturday morning and they were lounging on the lovely sofa eating bright red too-early strawberries in blue-and-white bowls. It was a perfect moment, in a very ordinary way. Pale March sunlight fell in stripes across the floor, and outside the window small birds dipped and swooped against a blue sky. It was a moment of close togetherness, and she wasn't nervous about venturing into this new terrain: it felt easy and natural, like stepping through a door into your own house.

She took another strawberry from the bowl and toyed with it. 'I've always thought I'd like three.'

Silence.

'Don't you think three is a good number?'

He said nothing, but as the conversation's meaning had become clear, his arm around her shoulders had turned dull and stiff, and she sensed her words hitting the air with a *ker-thunk*. Lead weights. Horror and Doom. She looked across the table at him, but his face didn't say much, just blankness, and then: 'I'm not having kids with you, Sky.'

She felt sick. In a small voice she asked him why.

'It's just not something I can do,' he said, and then, 'It's nothing personal.'

She blinked. There was no doubt that they had different definitions of personal. Outrage spread through her insides, about the meaning of the six months since Spain, about the barebacking. What exactly did he think was going on here? Did he know how babies weren't made? It was only through sheer luck that . . . She took a deep breath and squashed her feelings down, tried to be calm and controlled, be reasonable.

'Is it because of your background? Are you scared you'd be like your dad?'

He did the eye-rub. 'Oh, please,' he said. 'Let's not start with the armchair psychology.' He sat up all serious-businessy, like someone reading the news. 'Look, Sky. I'm just being straight with you. If you want children, you'll need to have them with someone else.'

'O-*kay*. I mean, I wasn't even . . . but even so, it's a bit . . .'

She wanted to shout: What have you been thinking when you come inside me with no protection? What level of headfuckery is this?

But she restrained her anger, made it small. Told herself she should try to understand. *Remember, Lilian, don't be grabby,* Mother had told her often, both verbally and non-verbally, when she was small enough that the word of her mother was eternal and unquestionable. *A good girl puts others first.* And somehow, despite her great efforts to escape Henfield and Mother, the fundamental belief that her own desires should never come first still nestled deep within her, very close to the core of her personality.

'Tell you what, let's not talk about it,' she suggested eventually, congratulating herself on finding a constructive solution to a problem that seemed altogether too confusing and difficult. They were young; they had time. 'Let's just . . . do what feels right.'

He had placed a strawberry in her belly button and was nibbling at it. 'Does this feel right?' he asked.

'It would feel better here.'

And that was it. They would do what felt right. And even later, when she was forced to rethink everything, absolutely everything, she would come back to that intuition of rightness, and tell herself that it must have meant something. *I want you in my life.* The workings of the universe are so impossibly mysterious, the outcome of any given decision so complex and far-reaching that, in the end, for all our ferocious desire to plan, calculate and optimise, intuition is really all we have.

7.

ALMOST a year into New Britain, and London was on the turn. All the grubby old derelict abandoned spaces – the factories, the warehouses, the docks and industrial plants – were being snapped up by men with well-cut suits and big ideas: arts venues, high-end flats. Formerly rebel artists were opening restaurants with celebrity chefs. Feminism was over, women having been liberated to wear tiny bikinis and clutch pints on the front covers of magazines. Everyone – except Dan who, due to his new status as a world-changing kingpin, wanted to stay 'off-grid' – now walked down the street having too-loud conversations into a mobile phone.

Ecstasy was on the wane – nobody took just one drug any more. Nights out no longer followed a collective and unified come-up and come-down: now each individual followed the zigzag of their own appetites as they snorted, popped and smoked whatever particular cocktail came their way. Raves didn't happen haphazardly in fields or under arches but in mega-clubs with listings and door policies and entry fees.

The Old Guard, the squat bunnies, the rave generation, were blinking and looking around and clocking that they had to either (a) knuckle down to life in the real world, or (b) start marketing and monetising the unreal world, quick-smart. Those in the (b) category opened clubs, or started festivals. Before long some would find themselves hiking their prices and putting up security fences. Some would grow rich, a few very rich, and they would congratulate themselves on their creativity and daring. And the world-changing impulse that had been the original spark of that creativity and daring would fade, until it existed only as a kind of nostalgia.

For Dan, however, the flame burned ever-brighter. Since the Encuentro in Spain last summer, he had progressed from general dogsbody into Mr Big. Meeting the international revolutionary avant-garde had given him a belly full of fire. He had a new mission: to persuade and cajole the world-changing group into yet again raising its sights. Their target should be, he told them, nothing less than a glorious global revolution.

'This is not just about our own backyard!' he had announced, at the Tuesday meeting, on his return. 'This is about the world economic system! It's about exposing and taking down the shadowy puppet masters of neo-liberalism!'

'Oh, okay,' said Bendy Aoife, with a shake and a shimmer of her glorious red hair. 'Sorry, but can you just explain what that is again?'

'Don't you know?' retorted Dan.

'I mean, yeah, in a way,' said Aoife. 'In a vague and kind of foggy way. But could you explain it more comprehensively and exactly, in simple terms?'

'Not right now,' said Dan, impatiently.

'We don't have time for that,' agreed the Jez-Gazzes. 'We have a glorious global revolution to organise.'

'Oh,' said Aoife, looking unhappy, looking confused. 'Okay.'

Dan informed them that the world-changing group was now officially aligned to the Anti-Imperialist Global Network, under the ultimate command of the *comandante*, and alongside all the organisations that had sent delegates to the Encuentro. The purpose of this international web, as the *comandante* had told them that night under the stars, was to coordinate a global campaign against the puppet masters of capitalism, otherwise known as the World Trade Organisation.

'What's that?' asked Aoife, again.

'It's a secretive meeting in which the money-men formulate the rules of global trade, behind closed doors,' explained Dan. 'They decide who can trade and on what terms. Their decisions are opaque and completely unaccountable. And those decisions perpetuate the oppression of the global proletariat.'

'Ah,' said Aoife, looking even less happy, even more confused. 'The proletariat. I see.'

'The ultimate destination is in sight!' cried the new fire-belly Dan. 'Now is the time to fight for a glorious and equal world! An end to poverty! Employment rights for all! No more exploitation by the developed world of its poorer neighbours in the south! Nothing less than a New World Order!'

By this point the representatives of the Old Guard – Aoife, Mouse, Rev – had gone very quiet. They were perhaps thinking wistfully about sunflowers and lentils and pointless joyful dancing in the street. They were perhaps recalling the unruly spirit that had led them all here, and wondering what had become of it. But Jez and the Gazzes were sitting up straight, on the edges of their wobbly community-hall chairs. This was exactly the kind of Historically Important confrontation with the status quo that they were signed up for.

'Count us in!' they cried excitedly. 'A New World Order! Glory to the International Revolution!'

And then they paused. 'But,' wondered Len, who, yes, was still present, hanging in there, Ken at his side, their beards slightly greyer, donkey-jackets more frayed at the seams, 'how are we going to do that? Isn't it a bit tricky to organise an international revolution, when our airmail letters will take several days if not weeks to arrive? How will we know what all the other global world-changing groups are doing, seeing as they're on the other side of the world?'

'Thank you, Len, an excellent question!' replied Dan, who had been paying more attention in the 'Constructing Global Networks' *seminário* than his girlfriend. 'You see, in order to advance the international revolution we will need to make use of this new thing called the internet.'

'What's that?' asked Aoife, who had no shame about asking questions, and really didn't give a shit if anyone thought she was dumb. The others in the room felt intensely grateful that she had this characteristic.

'It's a digital web of information,' said Dan. 'You can access it using a computer. Any computer. So, for example, anyone in the world can send you a message, and you will receive it immediately, wherever you are.'

Aoife wrinkled her goddess-like nose, trying hard to imagine this amazing and surely impossible thing. 'You mean like a postbox?' she said. 'An instant, portable postbox?'

'Hmm, yes, something like that,' said Dan. 'Let's say,

a cross between an instant portable postbox and an encyclopaedia.'

Everybody in the room tried hard and failed to imagine that.

Armed with their revolutionary fervour, Dan and the Jez-Gazzes had thrown themselves into planning a new annual celebration of international workers' day. On 1 May 1998, the Anti-Imperialist People's Network wanted to co-ordinate an international street party: simultaneous road occupations across sixty countries, from Buenos Aires to Barcelona, Sheffield to Sydney. The world-changing group masterminded the London offshoot, which took place in Trafalgar Square.

For the first time this was not just a party, but an explicitly political action with an official set of demands, including dropping third-world debt and introducing a universal living wage.

The whole atmosphere was different, too. In the run-up to it, the police were determined to intimidate and repress. They sent officers to wait outside the world-changing group's Tuesday meetings, filming people as they came and went. The old collective magic was replaced by something grim, tense and pressured.

At the same time, other dark elements sensed an

opportunity. That May Day saw the arrival of a new demographic: the Angry Men, not fluffy and arty like the Old Guard, not serious and political, like the Jez-Gazzes. They dressed in black and turned up late in the afternoon, pushing their way through the crowds towards the plate-glass windows of McDonalds and HSBC. The next day, it was photos of the Angry Men smashing the glass, shards falling around their heads like confetti, that dominated the news.

Within the world-changing group, the Angry Men divided opinion. The Old Guard, the fluffies, Rev and Aoife and Mouse, regarded them as a parasitical menace who did nothing to help create the parties and everything to take advantage of them for their own violent ends. But the fire-bellies, Dan and the Jez-Gazzes, felt the Angry Men were a useful weapon to be deployed judiciously to keep the police on their toes. They would point out that thirty thousand people had turned out, 100 per cent peacefully, for the legalise-marijuana demo in Brockwell Park the previous summer, and they'd got no media coverage. Nothing. Not a single news report, nothing in the papers. The whole thing might as well never have happened.

If we remove the threat of violence we're too easy to ignore.

This became the argument, week after week, month

after month. It was the issue of violence that would eventually cause the world-changing group to fall apart.

And what about Skylark? She had already stepped back from world-changing, and her once-rambunctious and rebellious life had become smaller, quieter, less trouble-seeking. Where she'd once stood up to the powers-that-be, now she made her way carefully through each day. Her future was at the play scheme. Suze had moved it into a temporary Portakabin while the new headquarters were being built.

Where once she'd flitted between squats and encampments, she now nested quietly in her little council flat. Sometimes she wondered where her world-changing passion had gone, the almighty anger and conviction that had propelled her out of Henfield and into the squats and encampments. In moments of confusion and frustration, she wondered whether Dan had somehow leached it out of her. But most of the time she was content with her smaller existence, relieved to be away from the increasingly bitter disagreements between the people she loved best. When she remembered the old triumphs, it was almost as though they had happened to someone else, some old friend she had lost track of along the way.

8.

As a consequence of the tensions within the world-changing scene, Rev and Skylark had drifted somewhat apart. They had been seeing each other less and less, not because of any overt ill-feeling, more an implicit sense of their priorities and interests having diverged. But he still invited her to the opening of his new exhibition, in November 1998. The card arrived in the post: a comic-book-style woman with lasers shooting from her nipples, standing astride the Westway. 'The Reverend presents: Waste Land'. It gave a date and the address of a peanut factory in the far reaches of west London.

She had been worried that she wouldn't be able to find the place, but coming out of the station into a dark and empty street, she saw it immediately: one of the buildings had a set of huge green tentacles squirming out of its roof. Lit up against the gloomy sky, it was a scene straight out of a Marvel comic, a city-terrorised-by-giant-radioactive-Kraken.

There was a metal shutter down over the entrance but it clattered open when she tugged it. She stepped into a narrow hallway, which opened out onto what would once have been the factory floor. It was dark, and glowing in the centre of the cavernous space was a vast tree, welded together from scrap metal, its branches reaching out across the ceiling, uplit in blue.

As she got closer, new details revealed themselves: whorls in the metal bark, a small possum-like creature peeking out from a hollow, red LEDs for its eyes. The air smelt of fireworks, and she remembered what excitement, what danger felt like. She had missed Rev a lot.

She called his name into the dark, but there was only an echo for an answer, so she made her way around the tree and through a door into the next room. As her eyes adjusted she made out an array of his machine beasts, bigger and grander than any she had seen before. A bird was strung from the ceiling, its wings fashioned from car parts, its beak open in a menacing squawk; a truck was parked in one corner of the space, with the ribs and vertebrae of a giant lizard.

In the centre she saw a battered Land Rover, with a dragon's neck and head rising from the bonnet. A woman in a crash helmet and mirrored goggles was sitting astride the beast, while an oil-streaked man in blue overalls was tinkering in the chassis.

'Okay, Ivy, try it again,' he called up, and she did something behind the dragon's ears, which sent a great plume of flame shooting out of its mouth.

Feet ringing out on the concrete, Skylark made her way past Ivy and the dragon and up to the second floor. It was emptier there, the full moon casting pale oblongs through the tall industrial windows. There was just one more beast, looming at her out of the dark: a hulking, ravaged horse striding forwards on long metal legs. Its body was the carcass of a military helicopter, smashed and turned upside down. Its head was the cockpit of a plane, the windows making blind black eyes.

She stood in its shadow for a minute, taking it in. The thing had a dark and uncompromising power.

'Ah, so you've met My Little Pony,' said a much-loved voice behind her. There he was, skinny and rakish as ever, dressed in a red striped suit with brass buttons and a ringmaster's hat, his pale face glowing in the moonlight. She gave him a big hug, feeling his ribs through the jacket.

'Glad you could come.'

'Well, Rupert,' she said, using the name she could use only when they were alone. 'When did you stop being a hippie lunch-out and become a proper artist?'

'I like to think I'm both.' He gave a theatrical little bow. 'But thank you. Art,' said Rev, warming to his theme, 'must always resist the system. Even if our new leaders build this great monstrosity on the South Bank, art must still get into the forgotten spaces, the car parks, the old factories. It must maintain its position outside the agreed rules and laws of the marketplace. This is what people fail to understand.' He paused, leaned forward. 'That art, that joy and fun and delight,' he whispered, 'are the most fucking revolutionary forces in the universe.' He straightened. 'Screw politics,' he added. Then gave her a mischievous grin. 'Wait until you see this, though. Come.'

He stalked off towards a corner and swung open a rusty door. She followed him up a set of stairs, which emerged onto the roof of the factory. They squeezed past the green tentacles on their scaffold frame. Around the edge was a perimeter rail, with a view out across the tops of the warehouses and beyond, to the lights of Shepherd's Bush Green. There, on the horizon, was the M41 motorway, flowing away into the distance: a vast, sodium orange river, glimmering with the headlights of a million cars.

They stood there together, watching it.

'World-changing was fun, back then,' he said.

'It was magic,' she said.

'But it couldn't last. It was always going to eat itself, eventually. That action back in May did me in. I don't know who I dislike most: the police, the Angry Men, or

the people in our movement who accommodate them.'
Rev didn't mention Dan, not by name. He took a deep
breath and sighed it out, mist curling from his mouth,
like smoke from the truck-dragon's. 'This country is
starting to feel like a too-small shoe,' he said.

'It's the same size as it always was, Rev.'

'Nah.' He shook his head. 'There used to be places
here where you could be free. Bit by bit they've locked
it all down. And people forget, so quickly, what freedom
feels like. The parameters shift, and it doesn't even occur
to anyone to question all the things that have been taken
away.'

'Perhaps people just move on,' she said. 'They want
stability, they want safety. Trying to fight the way things
are just burns you out in the end.'

'Not me,' he said. 'I need freedom. I need it like other
people need telly, or cars, or shopping centres.'

'I wonder where,' she asked, 'would be free enough
for you.'

He turned towards her, pale eyes glinting. 'Soon I'm
going to leave, Sky. Not just on my own – a few of us
have been talking about it. I'll do one more party, then
I'm off. We'll go in a convoy, travel across Europe, see
what we find. Squat some spaces, make some art.'

Her heart contracted, and tears prickled in her eyes.
She and Rev had come so far together. The sibling she

didn't have. He had made it all right for her to be a weirdo, opened up a life for her beyond Henfield, beyond McDonald's in Haywards Heath. What would have become of her if he hadn't been there to sneak off to the Traveller site, to run away to the woods? Would she still be in Henfield? Would she have developed strong feelings about Tupperware, about parking? It was impossible to imagine herself without Rupert, without Rev.

'You can come, you know,' he said. 'Come with us, if you want.'

'I can't,' she said. 'I've got a job, and a flat, and—'

'And a man,' he said. 'I know. You've settled in, hey? Well, I'll send you postcards.'

'You'd better.'

They stood for a moment, watching the motorway as it hummed and flowed. Then he leaned over, and whispered, very softly, 'Just promise me one thing, Skylark McCoy. You will never get *too Henfield*.'

And she whispered back, 'Never. I promise.'

9.

REV's last street party was the world-changing group's most improbable success, and also its worst failure. The plan for May Day 1999 – the most ambitious yet – was to put on a rave right in the heart of the City, the financial district. This was one of the most surveilled and policed square miles in the world, and attempting to take it over looked a lot like a suicide mission. It was almost unthinkable that it would succeed.

But, at first, it did. Ten thousand people congregated in Liverpool Street station, and following a brainwave of Rev's, they were each given a mask to wear in one of four different colours: red, blue, gold or pink. On the back of the mask was printed a set of instructions: *At the signal, find the flag that matches the colour of your mask. Follow your leader.*

This divided the crowd into four separate sections, each of which went off in a different direction. It was intended to confuse the police, and it worked: their command structure was not equipped to cope with four different directions of travel. Four different roads blocked; four different parties. In the resulting confusion one of the groups was left entirely unmonitored, and managed to get right inside one of the major stock exchanges.

But if the initial stages went better than could possibly have been expected, the day went on to go totally tits-up. The first mistake was fairly and absolutely squarely Dan's. His designated role had been to slow-crash one of the cars to block off Party Area Two. That parked car was integral to holding the space. But for some reason that nobody could understand he parked it in the planned spot, leaving the driver's-side window wide open. All the police had to do to remove it was reach in, take off the handbrake, and wheel it away. They retook the road and sent all the protesters from Party Area Two packing.

It was a simple mistake, but it wound up Rev no end. *You're full of the big talk*, Rev would rage at the debrief afterwards, *but you fuck it up with something so simple. How could you forget to close the window?*

Those two now openly maddened one another. Any attempt to be polite and respectful of each other's differing standpoints had evaporated.

I just forgot, Dan said lamely. *Sorry.*

But this was only the more minor of the day's fuck-ups. The major was that the world-changers had naively failed to factor in the aggressive alpha-male nature of the brokers inside the stock exchange, or how utterly contemptuous they would be of a bunch of scrawny, grubby anti-capitalists invading their workplace. Never in all its history had the group met with such a violent response. It was immediate, and it was merciless. The City boys raced off the trading floor and piled in, grabbing protesters, punching and kicking them and, in the case of two Gazzes, beating them senseless.

To make matters worse, as this ruckus was in progress the Angry Men turned up outside and again smashed up the local McDonald's.

The next morning all the papers – and not just the *Daily Record* – unanimously declared the world-changers to be violent thugs who terrorised kids as they ate their Happy Meals.

'They're just parasites,' Dan raged, pacing around the living room red-faced and agitated. 'They don't bother to help make it happen in any way. Just wait for us to create the space, and then charge in to mess it all up.'

Skylark shrugged up at him from the sofa. 'They're Angry Men, Dan. What do you think they're going to do? If you call for a revolution, idiots will turn up with baseball bats. That is the way of the world.'

'We didn't call for violence.'

'But you didn't call for non-violence. You left the door open.' She pulled him down next to her, kissed him, stroked his hair. 'And talking of leaving things open, what happened with that car?'

'Oh, don't you start,' he said.

'It's so unlike you. You never make mistakes.'

'Everyone makes mistakes,' he said. And he smiled at her so sadly that she let it go. He took a deep breath, settled back, calmer. 'Rev's leaving, isn't he?'

She nodded. 'But not just because of this. He's had one foot out of the door for months.'

He kept his eyes on her. 'And you?' he asked.

'What about me?'

'Do you want to go with him?'

'If I wanted to, I would.' He tugged her hand and she climbed onto his lap, pressed her body against his, felt him let something go. He buried his face in her neck. 'Do you want me to stay?'

'Why do you ask? Of course I do,' he said. He was lifting up her T-shirt, unclasping her bra, burying his face in her chest.

'We should use something this time,' she said. 'It's that time of the month.' He pushed her back, pinned her down with his weight. 'Dan? Did you hear me?'

'Mmm?' he said, but then, 'mmm.'

'Mmm?'

'Mmm.'

DI Wells: Well, well. That was dramatic.

UCO122: The wearies are very despondent about it.

DI Wells: I imagine they are. They just lost all public sympathy overnight. And if there was any doubt about the funding for this unit, that is well and truly put to bed now they've trashed a major stock exchange and terrorised a number of bystanders, including children.

UCO122: Quite. The group now looks likely to split. Several of the older members – the ones who are committed to non-violence – are leaving.

DI Wells: Where are they going?

UCO122: Aoife has been made an offer she couldn't refuse by a travelling circus. Mouse has received a holy call and is headed for a meditation cave in Tibet, or possibly Mongolia, I'm not sure. Rev – Rupert – is going to travel around Europe.

DI Wells: Allowing us to remove that car was quite audacious, I must say. Did you get much grief for it?

UCO122: A bit. But in general that minor mishap was overshadowed by the complete shit-show that followed.

DI Wells: But once again this year there were, you've probably seen, similar street parties in other cities, all over the world. It's quite remarkable. They've actually managed to construct a functioning international anti-capitalist network.

UCO122: Not bad going, when you consider their general . . . operating level.

DI Wells: How have you been getting on with the trainees?

UCO122: Well, I've passed on all the obvious stuff. They're very experienced operatives, having come from Northern Ireland. I think they may find themselves getting pretty bored.

DI Wells: So Tunbridge Wells isn't the hotbed of dissent we thought it was?

UCO122: You heard it here first.

But seriously, there are certain groups with a greater capability that do need monitoring. The anarchist guys particularly. I wonder why we're not doing that. Most of the outfits the new UCOs are being sent into are basically just a bit of banner-waving on the village green. It's a sledgehammer to crack a nut, if you ask me.

DI Wells: Word from the top is that domestic extremism is seen as the number-one threat to our national security. Your work in Spain and at home has been instrumental in convincing a lot of people in high places that this unit is a fundamental necessity in ensuring our national security. So you may take a great deal of credit.

UCO122: Thank you. Actually, Martin . . .

DI Wells: Yes?

UCO122: I have a question.

DI Wells: Of course, fire away.

UCO122: Is your feeling that I'm secure in this posting for the foreseeable future?

DI Wells: I haven't been told otherwise. Why?

[PAUSE]

UCO122: No particular reason.

DI Wells: Political events being the way they are, and having gained the position you're now in, I don't see the high-ups being in any hurry to pull you out. You're too valuable where you are.

UCO122: That's what I thought, but . . . Thank you, Martin. That's good to know.

10.

SHE knocked on the frosted-glass door of the corner office. 'Suze?' she called. 'I'm off now, if that's okay.'

No response. She knocked again, then opened the door and stuck her head around it. Across a wide stretch of green carpet, Suze was sitting in an electric-blue swivel chair, with her back to the door. She spun around, the phone clutched in one jewelled hand, gesturing to hang on a moment. Suze's dress sense had become even more extravagant – she was well known for it now, got recognised in the street. Her dresses were more voluminous, her scarves brighter-coloured, the rings and earrings gaudier than they had been when she was a humble everyday play-scheme manager.

Her clothes were part of the reason the press loved her, although Suze would never admit she cared about things like that. When the papers needed a quote about disadvantaged children or social inequality, she added an appealing splash of colour to the pages. And as her public persona had grown ever more vivid, the real person underneath became harder and harder to make out. She didn't spend much time with the kids any more, but sat in her office, making important calls, only popping out now and then to sweep around the centre, trailing perfume and doling out exuberant hugs.

Today she had on a turquoise robe with bright orange flowers, an orange headscarf and pink earrings.

'Well, that's very exciting, Neil darling,' she was saying into the receiver. 'Look forward to discussing it over lunch. See you next week.'

She hung up and sighed. 'That was the Tate's director. They want us to do an exhibition, with the kids, some kind of installation to highlight the impact of child poverty. I don't know about art, Sky. Do you think your friend what's-his-name – Vicar? – would help us?'

'Rev, you mean? He's away at the moment, out of the country,' she replied.

'When's he back? We could pay him, of course,' said Suze. 'There is money.'

There was always money now. The new centre had opened late last year, on the site of an old glassworks in east London. Tessa's friend had designed it, an architect who usually worked on art galleries and high-end restaurants. It was all space and light and white walls and primary colours; as well as a soft-play area with a ball pit, there was the hydrotherapy pool downstairs and also a computer room filled with Apple Macs, which nobody knew how to use. The chairs, which were in primary colours and shaped like flying saucers, had been donated by a famous interior designer. It was hard to work out whether to lean back in them, like an armchair, or sit upright: neither position felt quite right.

'Oh, please, do ask him, darling. They'll try to make me do one of these awful open recruitment processes but, as you know, I always prefer to work with people who are part of the family.' Even as The Crew had grown, Suze had maintained her resistance to bureaucracy.

'I will,' Skylark said, 'if I hear from him. But I'm off this afternoon, remember?'

'Oh, yes,' said Suze, vaguely. 'What is it – a doctor's appointment?'

'That's right.'

She hadn't told Suze why she was going to the hospital. It was too early to tell work about the pregnancy, particularly as Suze, who had never had kids, did not always take an enlightened attitude towards maternity leave. When one of the other managers had got pregnant, Suze had given her job unceremoniously and illegally to another more junior member of her team. *The young people here are our family. They rely on us. We can't get distracted.*

'You've had quite a few lately. But I won't ask.' Suze's black eyes glittered. She still didn't miss much.

Skylark left the office, heading out into the drizzly summer afternoon. August now, and it was thundery and humid, heat pressing in on her beneath thick grey clouds. Dan had been under pressure too, presiding over the fall-out from the action in the City. Rev, Aoife and Mouse had all left; the world-changing group was regularly monstered by the media. The stress had been such that she hadn't yet found the right moment to tell him about the mistake that was growing inside her. The urge to do so had been overwhelming, of course, when she first saw the two lines appear on the stick. But an instinct had held her back and it did not go away.

I'm not having kids with you, Sky.

What would he say? Would he want her to get rid of it? Every time she had opened her mouth to tell him, the words had stuck in the back of her throat. So she had trailed lackadaisically through the last twelve weeks, going to bed almost as soon as she got back from work and telling him that maybe she had the flu.

At the bus stop she got her mobile phone out of her bag. She thought about calling Dan. She thought about him coming with her to the scan, maybe, happy and excited like a normal dad. He could hold her hand and marvel at the miraculousness of the tiny mistake her body was creating. That together they could anticipate the new future – a real future, as a family. This was all possible. It was only in her own mind that she feared his response might be otherwise. But what was she actually waiting for? If she didn't tell him now, then when?

Heart pumping, she dialled the flat's landline and waited for him to pick up.

'Dan?'

'Oh, hey. Can I call you back? I'm just—'

'Well, actually, I—' She stopped.

'What? Is something wrong?'

'I need you to meet me at the hospital.'

'What? Why? Now?' He sounded irritated. 'What's going on?'

'I'll tell you when I see you. Just meet me at the Homerton, outside the gates. I'll be there in fifteen minutes.'

'But—'

'Just be there, please,' she said, and hung up.

She stood outside the gates watching buses pulling past, kids booting a ball in the front garden of the estate over the road. Well, it was too late now anyway: he couldn't make her think again. The mistake already had a spine and eyes and the bud of a brain, and she wasn't killing it, no way no how. He could say what he liked. Leave her if he liked. Too late: it was too late.

I'm not having kids with you, Sky.

He had told her, hadn't he? He had told her, and she hadn't listened. She had, in fact, completely and wilfully ignored him.

But then . . . he could have stopped. He could have taken more care. She had warned him. He was to blame for it as much as she.

She told herself this, had been telling herself this repeatedly for twelve weeks. But did she believe it, really, in her heart of hearts?

She did not. She did not believe it. Deep down, she believed that this pregnancy was her fault and her decision. She had wanted it. He had told her he didn't, in no uncertain terms, and in allowing and encouraging him to come inside her again and again, without questioning this contradiction, she had put herself at fault.

And now, when he realised what she had done, he would be angry with her and blame her and leave her. She took a deep breath, got ready.

He came around the corner, walking fast. Right from the other end of the street she recognised irritation in his hunched shoulders and his frown. She waved, making it light – *Keep it light, for God's sake keep it light* – as he drew close.

'So what is all this? Are you okay? I was just in the middle of—'

'I'm pregnant. Okay? I'm pregnant, Dan.'

She said it, quickly, before she could chicken out again. And then she had done it, she had said the words, and she would never again live in a world in which they had not been said. He stopped, shocked, double-took. He went very pale, and almost said something but stopped himself. Then he did say, 'Oh, my.'

For a moment they stood there, suspended between the knowledge and the reaction to the knowledge. For that moment, the situation just was what it was. Then a smile spread across his face and he gathered her into his arms, his face buried in her hair, warm breath in her ear. And then, after a moment, 'How long?'

Despite herself, she sobbed. Perhaps the relief of it. 'Twelve weeks,' she told him. 'It's the first scan today. That's why I'm here.'

'You've kept this a secret for twelve weeks?'

'You're not the only one who can keep a secret, you know.'

He stroked strands of hair away from her damp face. He kissed her tears. Relief flooded through her body, so she almost wanted to release all her muscles, fall to the ground.

'I've been so worried. Because you said you didn't want kids. I thought you might—' *Want me to kill it*, she thought, but did not say. *I thought you would want me to kill our child.*

'I'm – I'm—' He was still smiling, gesturing around him, as though to pluck suitable words out of the air. 'Well, it's amazing,' he said. 'It's—'

'What are you going to do, Dan?' she said. 'Will you stay and be a family?'

'Is that what you thought? You thought I'd leave?' He was holding both of her hands.

'I wasn't sure. I'm sorry. Sometimes I can't tell. I don't . . .' she said '. . . always understand you very well.' And with this she had to sit down on the low wall between the street and the estate, and he sat down next to her. 'So you're happy?'

The eye-rub. 'It's big news.'

'Did it maybe occur to you that this would happen,' she said, 'seeing as we haven't used contraception for so long?'

'I guess I haven't really been thinking about it.'

'How is that possible?'

'Oh,' he said sharply, 'so you *were* thinking about it, were you?'

She prickled at his new tone. 'Not like that,' she said. 'Not like I was trapping you into it. It just seemed a bit obvious, that's all, what the outcome would be in the end.'

She crossed her arms and bit her lip, staring intently at the hedge, the bus, the ball. He said nothing for a while, and as they sat there together, the air between them charged with something furious and ominous and unspoken. But when she looked back at him she saw his eyes were shining wet, and he reached out again and pulled her close. As they breathed in unison, whatever it was dissipated as quickly as it had come.

'I will be happy,' he said. 'But I'm— It's a big new thing. Just let me get used to it, okay?'

'Okay.' She looked at her watch. 'So are you going to come with me?' And then, when he didn't immediately understand, she said, 'The scan! You know, we can see the baby.'

So there he was, sitting next to her, holding her hand as the blue-smocked nurse slid the probe over her still-flat tummy. They watched together as the white pixels moved across the black screen, swirling and resolving themselves into a white cone-shape with a small, twitching black tadpole at the bottom. Through the monitor, a racing sound, a drumming of hoofs, the future, galloping towards them. 'There's the heartbeat,' said the nurse, and Dan's grip on her hand got tighter, too tight. As the nurse took some measurements they watched the tadpole squirm and flicker.

'Oh, my,' said Dan, again. 'Look at that. Okay.' He rubbed his eyes. 'Wow. Hello there.'

'Any feelings to report, Dan?' she asked mischievously.

'Possibly,' he said. 'Just a few.'

11.

'HEY,' said Dan, 'look at this.' He handed her Mother's old *Daily Record* (when visiting Henfield, he always read the old *Record*s Mother kept in her special wicker *Record*-holder, because *Know your enemy*). She took it and read the caption first. 'TANKS A BUNCH: A British artist known only as the Reverend recreated Stonehenge in Sarajevo, using vehicles abandoned after the war.' The picture above it covered half the page. Armoured cars – there was actually only one tank – stood on their ends in a circle, towering over a blasted landscape, weeds growing through heaps of rubble. Two leaned together to form a bullet-proof arch, framing the rising sun.

'An artist known only as The Reverend!' she said, delighted. 'He's famous!'

'Who's famous?' called Mother, from the kitchen. It was the last Christmas of the millenium, and she was buttering parsnips, while Skylark and Dan sat lazily reading papers at the fake-mahogany table, the baby photos topped with tinsel and the brass clock tick-tocking. The future was on everybody's minds, that Christmas. What would it hold? A complete breakdown of all computer systems, and a plunge back into the pre-internet dark ages? The ultimate triumph of the global anti-capitalist revolution? Almost certainly – barring some unthinkable twist of Fate – it would bring the birth of a new human, an amalgam of his-genes and her-genes, a small mistake who would bind the two of them there present together for ever, for better or for worse.

'It's our friend Rev,' explained Dan, even though Rev definitely wasn't his friend, not now. Rev had left back in June, leaving Dan to occupy the Throne. *He's welcome to it*, Rev had snarled to Skylark while somewhat the worse for wear at his farewell drinks. *Good riddance, frankly. I'm done. Over and out.*

As promised, Rev sent her a postcard every now and then, usually some kind of technicolour image of a Communist-era housing estate or triumphal golden statue of a dictator. On the back would be a stream-of-consciousness motivational message in his unruly scrawl. *It's ljovely ljubbly in Ljubljana! Imagining the world as it could be, wild and free . . . Keep singing my one and only Skylark.*

Mother came into the room, wiping her hands on her pinny, to peer over Dan's shoulder at the paper. She pushed her glasses down to the end of her nose and frowned. 'I don't get it,' she announced, after half a second. 'This is modern art, is it?'

'Apparently so,' replied Dan.

'You can just stop making your bed and call it art, these days,' said Mother, whose *Records* had kept her up-to-the-minute on the latest cultural developments. 'Or stick some rotten old fish in a tank.'

'Make yourself millions,' agreed Dan.

'What happened to real art, like paintings?' asked Mother, warming to her theme. 'You used to do lovely paintings when you were at school, Lilian.'

Skylark sat there, trying to contain her baby-heavy bad temper. When Dan and Mother got together they could exacerbate each other's most objectionable tendencies. In a world-changing setting he might have been all anti-authoritarian bluster, but on their visits to Henfield Dan was charming, deferential, respectful. He did things like opening doors for Mother, helping her with the computer she found a constant source of utter panic and confusion, putting up a new shed on her allotment. As soon as he set foot in Henfield, among the leylandii and crazy paving, something inside him seemed to relax. He rubbed his eyes less, smiled more.

For her, on the other hand, returning to Henfield meant regressing to some kind of monosyllabic teenage self that she liked to think she had grown out of. Spending time with Mother was an awkward and provoking reminder of things that still made her sad. She hadn't wanted to come here for Christmas; she'd tried to persuade Dan to contact Diane, his stepmum, so they could go up to Yorkshire and meet her and Sidney before the baby was born. But he wasn't having any of it.

'Not going to happen,' he said, in the same flat-out way he had said, *I'm not going to have kids with you.* 'I'm sorry, Sky, it's just not.' With this he ended the conversation, saying he had to leave for a Very Important Meeting, exiting the flat and closing the door firmly behind him. So that, it seemed, was the end of that.

Back in Henfield, Mother was still rhapsodising about her daughter's artistic prowess. 'I've got one of her paintings upstairs. Her art teacher told me she was one of the most talented in her year,' she was saying to Dan, and then sighed. 'If only she'd stayed on to get her A levels.'

'I'd love to see it,' Dan said, somewhat hurriedly, knowing that the conversation was edging towards one of the many flashpoints between mother and daughter.

'Of course,' said Mother. 'I'll go and get it.'

'Oh, please, leave it out, Mum,' Skylark said, in the ratty tone she reserved for Mother. But Mother took no notice, flitting off to retrieve her sixth-form painting.

She looked again, more carefully now, at the photograph in the paper, and filled with pride for Rev: he had realised a vision that she could now see he had been working towards for a long time. He'd always made art out of discarded things, teasing out creatures with souls from the junk people threw away. But the circle of standing cars took that to a new level. It was completely of this moment, but also ancient; it was war-like, but also peaceful. And it wasn't complicated or over-conceptual, evidenced by the editors of the *Daily Record* choosing it to entertain their readers.

Well done, artist-known-only-as-the-Reverend, she told him silently, beaming her thoughts out to wherever her old friend might be. *The arrow of hope. You've done it, just like you wanted to.*

'Here it is,' said Mother, bustling back into the room holding the small square canvas that usually hung above the never-used piano. It was an oil painting of Father, which not-yet-Skylark had done from an old photograph, just after he died. She remembered painting it, how for those hours she had been completely absorbed, feeling almost as though it wasn't her doing the painting at all, like her normal self was absent, and that instead some force outside her body was guiding her brush as it moved between the colours and the canvas. At the end she knew that whatever-it-was had caught something of him. It was a close-up of his face, but the palette was almost tropically bright, orange and yellow and green, revealing the unconventional spirit behind his features, his slightly hooded eyes, his bowl-cut hair.

'It looks just like him,' Mother said proudly. 'Honestly, if she'd just stuck at it, she could have been—'

'But I am,' she said quietly. 'If you stopped always thinking about what I could have been, perhaps you'd be able to see me, right here in front of you.'

Mother halted her babbling, but otherwise ignored this comment. Dan shifted awkwardly in his seat. 'It's great, Sky,' he said. 'You're talented.'

She was grateful, so grateful to have him there. With Dan at her side, it didn't seem so impossible to reconcile herself with Henfield, with her Mother-land. He rested his hand on her knee, as if to say, hang in there, it will be over soon, a solid buffer between herself and her past, this man with whom she was moving brightly into the soon-to-be-revealed future.

'Well, I thought it was absolutely shocking,' said Mother, once the plates of turkey-and-trimmings were half eaten, the crackers cracked, wine drunk. 'Honestly, it was like the end of the world.'

They were talking about the recent demonstrations that had taken place in Seattle, against the World Trade Organisation. Protest groups from across America had flooded into the city to disrupt its latest summit. They'd partied in the street; they'd had stilt walkers and fabric sunflowers and all the rest of it.

But it had turned bad. The Seattle police had gone in with tear gas and plastic bullets. The TV news showed footage of the streets filled with smoke from burning cars, cops dragging protesters across the tarmac by the hair, protesters lobbing bricks at the cops. The summit had had to be abandoned, but the scenes were ugly.

'Maybe it was the end of the world,' agreed Dan, whose lips were wine-stained, his cheeks flushed. 'The beginning of the end, at least.'

'All that smoke,' said Mother, still in her shocked voice. 'And people throwing things. All those hooligans chucking bricks at them. I mean,' she looked up at Dan, 'those poor police officers.'

Dan paused, a forkful of sprouts halfway to his mouth.

'In these things you organise,' Mother went on, 'these street demonstrations, you'd never advocate that kind of violence, would you?'

He inserted the sprouts and chewed, buying time.

'It's complicated,' said Skylark, before he could reply. 'Wouldn't you say, Dan, that it's a complicated issue?'

He put his fork down, swallowed. 'We have never advocated violence,' he said.

'But they haven't condemned it either,' Sky explained. 'They just kind of fudge the issue.'

'We do not fudge!' cried Dan. 'We are absolutely not a violent movement.'

'Good,' said Mother, vehemently. 'I know you mean well, Dan. I just couldn't bear to think you were one of Those People. You've never seemed like that to me.'

She glowed at him over the top of her wine glass.

'Oh, no,' said Dan, breezily, reaching for the carrots. 'I'm not one of Those People. I'm not like that at all.'

By the time dessert came around both Dan and Mother were well soused: between them, they had got through three bottles of red wine and made a good start on the port. Mother sloshed brandy over the pudding, and Dan got to his feet unsteadily to close the curtains and switch off the light.

'The big moment,' he said.

Mother lit a match and the thing went up in a whoosh of blue flame. She beamed broadly at the two of them, faces lit by the flickering blue.

'I would just like to say,' she said, slurring slightly, grasping Skylark's hand, 'how pleased I am. To see you two . . . together. You know, I worried . . .' she looked at Dan confidingly '. . . I worried so much about Lilian. We went through a lot, as a family, what with the tragic losses –' she swallowed '– of my darling first daughter Janie, before Lilian was born, and then the death of Lilian's father, to whom she was very close. I was doing my best' – tears gathered and brimmed over – 'in very difficult circumstances. To keep the show on the road, as it were. It's all very well to look back and think things could have been different.'

The flames on the Christmas pudding had guttered and gone out. They sat in the darkness as Mother's voice went on. 'So I can understand, I have always understood, why she ran away and went over to the wild side. It may not have seemed like it, Lilian, but I always knew why you did that, why you threw away your education and took off for the woods. I always knew it was my fault.'

'But, Mother,' Skylark said, 'don't you see? I didn't throw anything away. That was the best decision I ever made.'

But Mother wasn't listening, wrapped up in her own sorrowful narrative. 'You two will understand soon enough,' she said mournfully, 'the challenges of being a parent. You'll find out that, however hard you try, you will always want, more than anything, to do better.'

'"They fuck you up,"' intoned Dan.

'Oh,' cried Mother, 'oh, but they don't mean to! They try so very hard not to. And they punish themselves so badly when they fail. But in the end,' she was gripping her daughter's hand now, 'we're all, every single one of us, only human. We can't protect our children from the harshness of life.'

Now Skylark took Mother's hand in both of hers, looked her straight in the eyes. 'Mum, listen to me. I am my own person. I have made my own decisions. You did everything you possibly could for me. I know that. I wouldn't want anything to be different, nothing at all.'

Their mother-daughter bubble was popped by a strange squeak coming from the direction of Dan. They looked in surprise at him, to see his head down.

'Dan?' Skylark said, shocked. 'Are you okay?'

He wiped his eyes, looked up. 'Sorry,' he said. 'Yeah. Absolutely.'

'Are you crying?'

'No,' he said. 'Hiccups. Too much wine.'

Mother reached out and affectionately chucked his cheek. 'He's a softy, really, this one.' Her face glowed again, with the light of affection. 'I'm so glad,' she said, 'so very glad to have you here, Daniel. I want you to know how welcome you are. For a long time, we had a gap at the end of our table. But now you're here, and a new baby's on the way. For the first time, in so very long, I feel that we are . . . a whole family.'

Dan shifted uneasily in his seat. But Mother unrepentantly raised her glass. 'I propose a toast,' she said. 'Here's to you, Daniel Greene. Welcome into our family.'

III.

2000

8. WITHDRAWAL

[...]

8.5.1. One could pretend to have personal difficulties that cause you to leave the political arena.
You will have to think through circumstances
where such an approach could be effective –
[REDACTED] In any case the difficulties must
appear to ring true and are dependent on your
particular target group's politics and attitude to
members.

Special Demonstration Squad,
Tradecraft Manual

1.

*D*ANIEL *Keller.*

She turned over the driving licence in her hand. It wasn't like she'd searched for it – it was just there, in the glove compartment of the van, along with the cracked tape-boxes and tissues and tins of barley sugar. It was Dan, all right, in the photo: square jaw, dark eyes, gappy front teeth. The face she had seen when she'd opened her eyes every morning for the last four years; she knew it better than she knew her own.

But she didn't know the name.

Keller, it said. *Daniel Keller.*

On the other side of the windscreen, it was a dull grey day on the Holloway Road. The Christmas lights were still up, although it was the first week of January. None of those big dramatic maybes had come to pass: the world hadn't ended, the computers were still working, the cogs of global capital were still turning, cars were still belching petroleum fumes out into the atmosphere, chainsaws were still buzzing away in the rainforest. The whole rotten system had survived intact, as the clock ticked past midnight and a new millennium began.

But that name. It set her bones humming.

She stuffed it into her pocket. The world outside retreated into the far distance: the shoppers with their SALE bags, two pigeons fighting over a box of fried chicken, the number 30 huffing past the stop. She was in a lonely faraway place, the icy top of a Himalayan mountain. She put her hand on her balloon-tight tummy and ran it along the right-hand side, where she could feel the curve of his spine. Three weeks until she would meet her baby. I'll ask your father later, she told him, told her humming bones.

So, when Dan came out of Mothercare with the buggy in its box, looking all satisfied with his new purchase, when he loaded it into the back and hopped into the driver's side, she didn't say a word. And all the way home, while he chatted away about *We need to buy peas* and *Did I tell you I'm away this weekend?* and *Remember to call the hospital, won't you?*, she stayed there on top of her Himalayan mountain and thought, *KellerKellerKeller*, repeated it in her head until it sounded like a foreign language. They bumped back down Drayton Park, past the Arsenal and down Seven

Sisters Road. The whole time one question was turning in her head.

Who is Daniel Keller?

She wanted to wait for a good, quiet, loving moment to bring that question out in front of him, to display it sensitively, like a delicate plate or flower arrangement. But she was not a patient, delicate or flower-arranging kind of human. The question blurted out as soon as they got home.

'Who is this KellerKellerKeller?'

She pulled the thing out of her pocket and his face did something she had never seen before. You know that painting *The Scream*? It did that.

'Where—' he stuttered.

'In the glove compartment,' she said. 'I was looking for a tissue.'

'Fucksake, Sky. What you doing going through my stuff?'

He snatched it off her and stumped into the kitchen. She stayed put and listened to him clatter around. Dan never shouted, never ever, but he could make washing-up sound like *Fuck you and everything you stand for*. He channelled a lot into his washing-up. After a few minutes she went in and sat down at the kitchen table and waited until he'd cleaned everything, including the sideboard and the salt 'n' pepper shakers and the washing-up rack itself, and by that time his face had returned to normal.

'I just met this guy who sold IDs,' he said. 'And I thought it might come in useful.'

'So it's a fake.'

'Of course it's a fucking fake, Sky. What else is it going to be?'

She had no answer to that. No alternative explanation as to why her partner of four years would have a driving licence in a name she didn't recognise. What she did know was that Dan was a world-changing kingpin, with his fingers in many pies. To the police, corporate fat-cats, judges, government and other representatives of the rotten neo-liberal democracy, he was a person of considerable interest.

There were reasons why he might want a fake ID.

Now he was laughing. 'What did you think? Like I'm a killer on the run or something? The look on your face.'

She allowed the relief to flood in, to silence the bone-hum. The world could go back to its non-faraway distance. She could come down off that cold, lonely Himalayan mountain and find herself back in Heron Court. Reality didn't have to realign itself today, not today when her tummy was huge and tight and her back hurt and she was having a baby in three weeks and really didn't have the energy for any kind of fundamental questioning of her daily existence.

Even when he stopped laughing and said, 'Just please don't mention it to anyone, okay?' it didn't bother her.

She just said, 'You knob-head,' and shoved his shoulder.

'Takes one to know one,' he said.

2.

THOSE were happy days, days filled with excitement, fear, anticipation and the heightened emotions parents-to-be have felt since the dawn of time. They had done all the normal things: painted the spare room, bought bottles and nappies. They went to birthing classes with a midwife called Helga, who lived in Stoke Newington. Helga was a hippie of the severe German variety. She made them lie down on her living-room floor and moo as loudly as they could. All the other dads were inhibited about this – they kind of tried to moo in a whisper. But Dan followed her instructions with full gusto.

'MOOOOOO.'

Oh, it did make Skylark laugh. She had to turn around and poke him and whisper, 'Stop it, you idiot, you're too much.'

'What? This is what she said to do, isn't it? Breathe from the diaphragm.'

'Zat's very good,' said Helga, who was standing on a chair, surveying them all spread-eagled in front of her. 'Dads, you must make zese noises too, make zem wiz all your hearts. Don' feel silly, don' feel shy. Ze vibrations of your voices vill give your partner great strength in managing her pain.'

That night as they walked home they discussed the matter of childbirth. Helga had been telling them about pethidine and forceps, contractions and epidurals, and the whole thing had freaked Skylark right out. She couldn't believe any of it would actually happen to her. 'I might give it a miss,' she joked. 'Keep him in there.'

Dan looked down at the bump, where it was hidden under her jumper. He'd been inspecting its progress every morning, watching it grow smooth and white, an alien egg. 'Careful what you wish for,' he said. 'You don't want to be two people for the rest of your life, now, do you?'

She considered this. 'I don't know,' she said. 'I'm two people at the moment, and it feels all right. It's company.'

'Believe me,' he said, 'that would wear off.'

Much later, she would remember that conversation, and see beneath it subterranean faultlines, great cracks and earthquake-ridden chasms running from their two small fragile human bodies right up to the highest eche-lons of the great capitalist-democratic institutions of the state.

That small exchange was a window into a world in which none of them present – not her, not innocent unborn baby, and not Dan – could truly be said to own their individual selves at all.

But she didn't think much of it at the time.

'In ze end, childbirth is just a few hours of your life,' Dan went on, in a Helga voice. 'Just remember zat, vatever happens, it vill soon be over.'

With this he reassured her, as he often did. Over four years, Dan's calm steadiness had slowly but surely transformed her life. It had given her a firm base, a steadfastness. A confidence about her place in the world – at least when he was around. The deep dark sea was still there, inside, but he kept her afloat.

'You're right,' she said. 'And you'll help me, won't you? The best moo-er in town.'

'Vhy, zank you.'

'Tell me it's going to be okay.'

'Of course it is, crazy lady,' he told her, losing the accent. 'It's going to be just fine.'

3.

Accepting his explanation about the pub ID man, she packed up the whole *KellerKellerKeller* thing and put it away in a dusty corner of her mind. At this point her mind was brimmingly full of normal things: tying up loose ends before maternity leave, paying bills, imminent childbirth. And also not-normal things, because those days before you have your first child are really not-normal. She was a landlubber about to embark in a ship on a long journey to faraway unknown lands. Mentally, she was packing her trunk, trying to anticipate what she would need to cross into a new continent.

Nevertheless, on the first morning of her maternity leave, for reasons she couldn't entirely explain, she searched his cupboard. This was a cupboard she rarely opened, as it was Dan's domain, a small territory in their shared bedroom that he had claimed for himself. It was where he kept his beer-brewing kits and unwashed sports socks and assorted dumbbells, sweat bands, half-used packets of Golden Virginia, train tickets roached at the corners, small piles of mismatched screws and Rawlplugs and drill bits. It was a man-cupboard for his man-junk, and ordinarily she gave it a wide berth.

But this particular day after he'd gone to work she was pacing around the flat and her bones had started up again their annoying and persistent hum. Without ever forming a logical argument for doing so, she opened the cupboard in the bedroom and pulled down the yellow plastic box from the top shelf. She took off the lid and scrabbled inside. She needed something to quieten the noise inside her body. Questions were bubbling up to disturb her brain surface, such as, how does a man live for thirty-five years without accumulating anything? Where were his bank statements, payslips? His childhood photographs and memorabilia? His cutlery, old mugs and books?

At first, the contents of the box seemed to be much of the same old baccy-filters-paper jumble. There were a couple of recent-ish photos, Dan somewhere hot-looking, in shorts and a too-bright Hawaiian shirt. A birthday card from her, with a soppy poem inside, in which she had rhymed 'Greene' with 'love machine'. She was readying herself to put the whole thing back on the shelf and close the door for ever when something caught her eye. There, right at the bottom, was a crisp passport, and a thick brown envelope, and inside the envelope his birth certificate.

And there it incontrovertibly was, in official black scratchy fountain pen on the line labelled 'Name, if any': Daniel Henry Greene. Date of birth: 14 October 1969. The rest was equally to-be-expected. Registration District: Barnsley. Name and surname of father: Sidney Walter Greene. Name and maiden surname of mother: Dorothy Greene formerly Ellis. Occupation of father: Coalface electrician.

In every detail, it was just as he had always told her.

Stop, stop that ungodly humming, she told her bones. Look at the evidence. A birth certificate, no less. Birth certificates don't lie.

Did that stop the hum? It did. Well, paused it, at least.

She took one more look at the birth certificate, all in order. Sigh of relief, followed by a dim confusion as to why relief should be necessary. Then she put the box contents back in their customary position, put the white plastic lid back on, shoved the whole thing into the man-junk cupboard, and shut the door. As she did so she banished all inconvenient and troubling doubts from her mind, until the day their baby nearly died.

4.

THE pain and its accompanying fear came on when she was walking back from the hospital, after having the last scan before the imminent birth. They were always strange, tense affairs, those scans, supposedly a heart-warming glimpse of her soon-to-be-child, but in reality a terrifying potential exposé of faults and failings, of disability and death. Having developed in recent years an anxious disposition, she dreaded the scan appoint-ments. That morning she'd begged Dan to come with her as she couldn't face going alone.

'Surely they'd give you the day off for this,' she'd said, before he left. 'It's the last scan before, you know, before . . .'

If he heard the doomy note in her voice, though, he didn't let it bother him. 'You know I can't do that, love,' he said. 'We're running behind on site.' And with that he gulped his coffee. But he must have glimpsed some-thing sullen about her expression, as he leaned over, kissed her, and said, 'I'm saving up my days off so I can be around for you both afterwards. I thought that's what you wanted.'

He was right. That arrangement was what they had discussed, and what they had decided. So she told herself firmly that she was being unreasonable and cranked out a smile. He was going away for three days, some building site up near Manchester. Such absences had always been a fact of life with Dan, and it had never remotely bothered her before. In fact, she had always relished the time she got to spend alone in the flat. He was handsomely paid for his aerial-access work, so handsomely paid, in fact, that he could cover most of their rent, and still devote several days a week to his world-changing activities.

But since becoming pregnant she had found his absences harder. On the days when Dan was not around, she felt bereft and abandoned. She knew she shouldn't take his going personally, imagining that he skipped out of the door forgetting all about her and the bump that would become their baby, but for some stubborn inner reason that *was* what she imagined.

Every time he left now, she felt – irrationally, she knew, completely irrationally – betrayed.

So that afternoon in early February she'd been at the hospital alone, the cool dry hands of the midwife squeezing her belly, like a ripe peach, the cold shock of gel, then the smooth, sinister glide of the scanner.

'Well, well, he's a big chap,' commented the midwife. 'Is his father tall?'

She didn't reply, trying not to think about the empty chair next to the bed, and Dan dangling from a rope somewhere very much not here. Everything, the midwife told her, was fine. The baby had turned, and the thing that felt like a foot sticking into her ribs was, sure enough, a foot sticking into her ribs. The head, according to this midwife, was fully engaged.

'Any day now, dear,' she said.

And that was what was echoing through her head as she made her way up Mare Street towards home. *Anydaynow, anydaynow*. Where would Dan be when *anyday* became *now*? Would he be with her, or far away, occupied with more important, significant and gratifying man-work? As these thoughts circled, as she approached the round church that marked the halfway point between hospital and home, she was ambushed by a stab of pain so sharp that she winced and held her breath.

That stab didn't feel right, not at all. Hippie Helga had told them zat ze contractions vould feel like a long, slow squeeze, the tightening of an iron band around ze area of ze vomb. This pain wasn't that. It was sharp, like a knife sliding into her liver. The first lasted only a second, but a couple of minutes later, coming up to Clapton Roundabout, there was another, this time so sharp that she cried out and clutched herself, doubling over.

When she opened her eyes the face of an old unknown Turkish man swam into view, concern in his silvery aged features. 'Madam?' he enquired. 'Are you in pain?' She still couldn't talk, which pretty much answered his question. 'Is there anyone you can call?'

She found that she could breathe again. 'Yes,' she panted, because there was, wasn't there? 'I'll call my husband. I'm nearly home. Thank you.'

(He wasn't her husband, of course: it was just easier to say that.)

And that was what she did, as soon as she got back, leaning heavily on the Turkish man's arm and bidding him goodbye on the front steps of their estate: she dialled the number that Dan had given her for the site somewhere near Manchester where he was doing his aerial-access work. He had always made clear that she should only call in emergencies, as he was usually a hundred feet up dangling from a rope and not exactly free for a chat. Calling on workdays in the aerial-access sector was very much not done, he had implied, and as a result she had never done it before (he persisted in refusing a mobile phone). On this occasion, however, she dismissed the non-calling rule without a second thought. *Anydaynow*, the midwife had said, and this pain was all wrong – if this wasn't an emergency, she told herself firmly, what was?

The number you have dialled has not been recognised.

The number you have dialled has not been recognised.

She hung up and dialled again.

The number you have dialled has not been recognised . . .

At that point the knife slipped back into her liver and the next thing she remembered was waking up in what was, if not an actual pool, then certainly a significant puddle of blood.

5.

Louis – that was his name, their son – didn't die, though. Despite her fuggy-headed weakness after the haemorrhage she managed to reach for the phone and call an ambulance. It was a matter of minutes before paramedic with a squashed-bun face was sitting on the side of her bed, voice warm as a blanket.

'Your partner is away?' he asked, and she replied, 'Yes, getting back tomorrow.'

The pulse machine beeped and whirred. 'Can you call him?'

'I've tried.'

'You've lost a lot of blood. I'll have to admit you. There's nobody else you can call now?'

There was nothing for it. She tried the emergency number one more time.

The number you have dialled has not been recognised.

A dark cloudscape massed in her brain.

How could he have given her the wrong emergency number?

Was it possible that he made a mistake when he was writing it down? But how could he have been so careless, when he knew that the due date was *anydaynow*?

The number you have dialled has not been recognised.

She put the phone slowly, deliberately back on the hook.

Inside, by now, she knew. Or, at least, she deep-knew, but couldn't allow this deep-knowing anywhere near the level where it would become ordinary, day-to-day knowing. She simply couldn't allow it, as she hung on desperately to the haemorrhaging pregnancy of his unborn child, to Louis, her already beloved son. What on earth kind of choice did she have?

There was no time for thinking, no time for anything. She reached for the phone and called Mother.

Later that evening, when the immediate medical emergency had abated, Mother had some Serious Questions. They were in the labour ward (the labour ward!), all lino and scratchy sheets and thin blue curtains and crying, both the endless wailing of infants and also the tears of their exhausted female birthers. Usually the noise would have set her teeth on edge but that night all it prompted was longing, a desperate longing that one day soon she would have one of her own, even if it did cry, even if it never stopped crying, to mean that all the pain, blood and fear wasn't for nothing.

'So where is he?' asked Mother, her lips puckered. 'Off on one of his world-changing jaunts?'

'No, Mother,' she replied, ratty with fatigue, blood loss, and also because that was the way she was with Mother, especially when Mother asked her to think about things she didn't want to think about. 'He's working. You know what it's like. He can't just drop everything and answer the phone.'

'I jolly well think he can drop everything and answer the phone,' said Mother, 'to his not-quite-wife who's about to put her life on the line to bring his child into the world.'

'Okay, okay,' she snapped, because now, in addition to reminding her that Dan wasn't here (and that they were not-quite-married, one of Mother's favourite topics), Mother was reminding her about her life being on the line, another thing she was trying not to think about. That was why she had wanted Dan, rather than Mother, to be present in her hour of need.

'He's back when?' Mother went on, regardless.

'Tomorrow.' She sighed, leaning back on the adjustable bed that she hadn't quite got at the right angle. Something felt odd inside the bump, like a blown light bulb, the filament tinkling free. Between her legs, blood was still trickling into the special jumbo sanitary pad provided by the nurse. It was an unpleasant damp out-of-control feeling, adding to her fear of moving at all.

'Total bed rest,' the doctor had said, a chestnut-skinned doctor with neat dreads and a firm smile. 'We'll need to go for an induction, but let's discuss that tomorrow once you've had some rest.'

'Tomorrow,' said Mother. 'Well, I shall be having a word.'

'Believe me, there's no point,' she said. 'He doesn't listen.'

'We'll see about that,' said Mother, who tended to believe that she enjoyed some kind of leverage with Dan that her daughter lacked. Usually Skylark found it maddening, but she didn't have the will to get annoyed as she lay there on the scratchy blue sheets behind the thin blue curtain.

'Mum,' she said, 'please shut up and just hold my hand.'

Mother knew when enough was enough, so she did just that.

6.

'I'M going to try the phone at the flat, just one more time,' whispered Mother, once Skylark had been hooked up via intravenous drip to a number of chemicals that had never been so much as mentioned by Hippie Helga during the birthing information sessions. 'If he picks up now he could still get here just in time.'

When Mother came back her eyes were wet. 'He was there,' she said. 'He'd just walked in the door when I spoke to him. Darling, he's on his way. Okay? He's coming. Just wait. He'll be here.'

And with that news she finally cried.

As soon as Dan arrived in the room, red-faced and panting, still in his work clothes, all the shadowy imaginings that had occupied her mind for the previous twenty-four hours disappeared in a puff of gas and air: she had energy only for now, for that precise moment.

He kissed her forehead and squeezed her hand and she could see he was shocked to find her in the hospital bed, and that he was sorry, he really was. Mother wished them luck, kissed them and left, and Dan took up his rightful position, at her side.

He rubbed her back through every contraction, and each one damn-near ripped her soul from her body, thanks to the non-hippie induction drugs. Each one was like climbing that Himalayan mountain, and when she got to the top, she knew she had to keep right on and climb it again. And that was what she did, for the next twelve hours, up and down mountains. She wasn't in the hospital at all, she was inside herself, climbing and coming down, climbing and coming down. Every so often Dan would pass her a cup of water or a Lucozade sweet, and she would swig it or nibble it, then carry on. Those refreshments and the pressure of his hands on her back were her only link to the world.

'You have no idea,' he said, stopping for a second and looking at his reddened fingers, 'how painful childbirth is for a man.'

As time went on, snatches of conversation made their way to her from a great distance: can't you give her something for the pain? Look at her, it can't be long now.

She was on a high, narrow bed, with stirrups for her feet. She had no idea how she had got into this new bed, who had taken her clothes off, or why, or what had happened to them. Something blue and triangular loomed into her field of vision: a woman, her face framed by a bright blue nun's habit. Her lips were moving.

'Take a deep breath, now.'

There were hands between her legs, and then a sharp stab and she lost herself in a white light of pain.

Louis was born at 4.14 a.m, on 6 February 2000. This scrawny purple-faced jet-haired creature slipped out like a fish and the blue nun lifted him straight onto her stomach. She recognised the touch of him, his tiny sharp-nailed fingers clawing her belly, and then she cradled him and met his startled, staring blue-black eyes. His first look was somehow a question, and from that moment on, her life would be devoted to answering it.

The world shrank around the three of them. He was white and exhausted, and he kept laughing, little breathy marvelling laughs. They murmured to each other, stupid things, the things new parents have always said. Look at his hair! Wait, just count his toes again . . . Will his head stay that shape? He's a weird colour, isn't he, kind of mauve?

The blue nun busied herself mopping and measuring and weighing and making notes in a little red book.

Skylark held Louis, whose eyes were closed now, and she could feel why he needed to shut out this too-much world – feel it as if his need was her own.

That was what struck her most about becoming a mum, the sensation that she would never be just herself again, that a piece of her would always be someone else.

DI Wells: Where were you yesterday?

UCO122: Yeah. Sorry about that.

DI Wells: Well?

UCO122: Something came up.

DI Wells: Please call, in that case.

UCO122: I couldn't. It was a – medical emergency.

DI Wells: Everything all right?

UCO122: Yes. Yes, thanks. Everything is fine. Everything is . . . good.

DI Wells: You look exhausted.

UCO122: Do I? Well, yes. I am. It's been quite a night.

DI Wells: So . . .

UCO122: So?

DI Wells: Do you have an update?

UCO122: Right. Yes. An update.

DI Wells: [SIGHS] Dear me, Daniel. This is our briefing. Remember? Let me give you a hint: this is the bit where you get to do your job.

UCO122: I do my job every bloody minute of every day, Martin. My life is my job.

[PAUSE]

DI Wells: Daniel, as you know, we encourage a flat structure in this unit. None of the cap-doffing you find elsewhere in the force. But I am still your commanding officer. Please watch your tone.

UCO122: [LAUGHS] Aye aye, Cap'n. Martin. Captain Martin.

DI Wells: So. What's the latest on May Day this year? I take it they're planning their annual shindig?

UCO122: Yes. I've made some progress. I recently made contact with a guy known as Anarchist Ed. One of the most wearisome wearies you could ever wish to meet.

DI Wells: I take that seriously, coming from you.

UCO122: Ed is one of the guys who co-ordinates the Angry Men, otherwise known as the black bloc.

DI Wells: Who are?

UCO122: The thugs dressed in black who turned up at the May Day protest last year and trashed the McDonald's. These guys really are nasty bits of work. They do nothing to help organise the actions, as they have a minimal political agenda. They're basically just goons who rock up on the day and commit random acts of violence. This group is the one we need to watch.

DI Wells: Noted. Thanks.

UCO122: Anyway, I managed to buddy up to Ed – helped him move his stuff into a new flat, the van never fails me – and he invited me along to some pub where they all get together.

DC2004: Right.

[PAUSE]

I'd like you to sit that one out, actually.

UCO122: I'm sorry?

DC2004: Orders are for you to sit that one out. Do not attend the black bloc meeting.

[PAUSE]

UCO122: Could you explain that decision, Martin? I've got access to some of the most violent elements of what will potentially be the biggest anti-capitalism protest that London has ever seen, and you're telling me not to go?

DC2004: That's correct.

[PAUSE]

UCO122: You do know that this could be worse than Seattle?

DC2004:   We're aware of the potential scale of this event, and I take your view into account, but those are your orders.

[SILENCE]

Daniel, where were you last night?

[SILENCE]

Were you, by any chance, with a certain weary, one who has a nice—

UCO122:   That's personal, Martin. It has no bearing on my work.

DI Wells: I'm afraid I disagree.

[SILENCE]

7.

SHE just couldn't get the hang of the feeding. When she put Louis's lips to her breast, he would open them just a tiny bit, not wide enough to suck. He did this again and again, for hours, after Dan had gone home to get some sleep.

It was two in the morning. The curtains had been drawn around the bed next to hers, and a posh male voice boomed out from behind them.

'What you have there,' this voice announced, loudly enough to wake up the whole ward, 'is a wound.'

The hospital was so hot and airless. Sweat trickled down the back of her neck. Her body felt battered, and between her legs the pain drip-dripped. The woman behind the curtains, the one with the wound, replied in a low murmur.

Exhausted, panicking, Skylark tried whispering into Louis's tiny ear. 'Come on, little 'un. Please. You know me. Skylark. Your mum, boy-o.'

But no way, no how. His mouth stayed shut. He kept smacking his tiny lips. They looked dry, they looked parched. Could a newborn baby die from dehydration? She wished she had read more books, listened to Helga more carefully.

'Has baby fed yet, Mum?'

She blinked. It seemed to be morning, and there was a nurse standing at the end of her bed. So many nurses, she never saw the same one twice. This one was young, petite, with an expression of something between mild concern and total boredom.

She shook her head.

'Show me.'

The nurse sat down on the blue chair. Louis's eyes were again closed. Panic rising, she brushed her nipple against his lips, and the nurse tutted.

'No, no, no. Look – like this.'

The nurse grabbed the baby's chin and pulled his mouth open, pinching her breast with her other hand. Louis cried out, a heartbreaking, quavering cry. 'You've got to teach him, Mum. You're in charge.' The nurse stood up, smoothed down her dark blue smock. 'You know I can't let you go home until feeding is established. And the ward is very busy today. Lots of people need beds. If he won't take the breast you can give him a bottle.'

'He won't feed,' she told Dan, when he arrived at eleven, rested and showered.

('Why can't you get here at nine, when it opens? Please, I really need you,' she had begged, but he'd said sorry, he had just one really important early meeting. 'A meeting? Today?' She was shocked, disbelieving.

'I really can't get out of it. I'll be there as soon as I can. Promise.')

By now she was fraying. She hadn't slept; her skin felt like raw flesh. She could hardly talk without crying. 'The nurse said that if he doesn't do it soon I'll have to give him a bottle.'

'What? That's rubbish.'

He didn't hesitate. Whenever she shrank and doubted, his confidence held firm. He believed in her, believed she could do this, that she could be a mother, even when she didn't. She felt a rush of gratitude towards him, her Practical Dan. Where would she be, without him?

'Let's just take our time. Maybe he's a bit stressed with all the moving around.' He stripped off his shirt and lifted Louis onto his chest, floppy head cradled in one solid hand. Louis curled there peaceful, ear pressed to his father's heart. She slipped into an exhausted sleep, waking up to find Dan placing Louis gently in her arms, then shaking him ever so softly to wake him up.

'Look,' said Dan, rearranging her. 'You need to line his

nose up with your nipple. Hold it, pinch it, yes, like that. Then wait until he opens his mouth and there!'

It didn't work.

'No, okay, don't panic. Just try again. There!'

With a little click, the baby slotted into place, and she felt a suction so strong it was hard to believe that the source of it was his miniature mouth. Louis's chin worked up and down, his eyes closed, like he was finally here, he had finally arrived. Her breast came alive with a tingling rush.

'Danny boy! It's working!'

'Of course it's working, Sky. That's what your body was made for.'

He was grinning away, and she smiled at him in relief. She had never needed another person this much. There was no way she could do this alone.

'But how did you know . . .?'

He got his *Daily Record* out of his bag. (*Know your enemy*.) 'Let's just say I've been around some babies in my time.'

She kept her eyes on him, as she said, 'Have you done this before, or something?'

He laughed, and buried his nose in the sports pages.

8.

THE first time she saw the buzz-cut men, she and Dan were on their way to the park, with Louis in the buggy. The buzz-cut men were standing outside Hamdi's newsagent on the high street. Two of them, big guys, close-cropped hair, dark jeans, tight T-shirts, standing in a wide-legged, arms-crossed stance. They might as well have had 'POLICE' tattooed across their foreheads, it was that obvious. When they saw Dan, one nudged the other and nodded.

Dan saw them, too. He caught her eye, and quickly looked away.

'Who are they?' she whispered, as they hurried on, but he just kept walking. 'Dan?' He didn't turn. 'What were those men doing?'

'Fuck knows,' he snapped. But then he said, more gently, 'Really. I have no idea.'

They walked to the park in silence. The bone-hum had started up again, still quiet but definitely there. It was a lovely sunny March afternoon, fresh, cold, the scent of spring on the breeze. The trees on London Fields had the slight aura of green that signifies buds on the way. For some reason, that made her want to cry.

In the period just after Louis was born, lots of things made her want to cry: baby animals, broken flowers, missed appointments, the hole in the ozone layer. He was so perfect, and the world wasn't perfect enough for him; she wasn't perfect enough for him. This simple fact broke her heart several times a day.

In the buggy, Louis was wriggling and kicking and starting to grizzle.

'Shall we sit here?'

'Sure.'

She settled down on a bench with the baby at her breast. 'So. Are you going to tell me?'

'Tell you what?'

'Why you're being watched.'

'We're all being watched,' he said. 'They're properly worried about May Day, after Seattle. They think London could go the same way.'

She had been vaguely aware, through the fog of breast-feeding and nappy-changing, that the police had once again stepped up their harassment of the world-changing group. Houses were raided, computers confiscated. The guerrilla gardening crew had been so badly harassed outside their team meeting that they'd had to abandon a bulk order of compost on the pavement. The cops had raided a circus-skills training session, and confiscated the trapeze.

After last May Day's chaos, after the full-on flounce-out of the Old Guard, tensions within the world-changing group had intensified. Two factions had solidified and now glared at each other every Tuesday night from opposite sides of the community hall. Those of a more pacific and hippie persuasion were determined to commit publicly to non-violence for the next action. They pointed out that while most ordinary people did not want to live next to a motorway, they also didn't want to be terrorised while peacefully satisfying their craving for sugar and saturated fats. *If you lose those ordinary fast-food eaters,* they said, *you've lost the argument. It doesn't matter how morally right you are.*

But the Jez-Gazzes had held out, and in the end, they reached a compromise. While there would be no public declaration of non-violence, this year's May Day street party would be packaged and promoted as a return to the carefree, arrow-of-hope spirit of the early days. The flyers would feature a big sunflower and the slogan 'This Is Not a Protest'. It would be all about workshops: guerrilla gardening and laughter yoga. With enough wholesome and diverting activities on offer, the argument went, nobody would bother rioting.

But this had not done much to convince the police to lay off.

'They're making pre-emptive arrests now. We've been classified as "domestic extremists".'

Dan got out the sandwiches she had made and poured tea and milk into the lid of the Thermos. They were looking out over the cricket pitch, its scrubby grass scattered with fag-butts. By the bins, a squirrel was feasting on a scattered packet of Wotsits. The dappled green light – or perhaps her woozy new-mum eyes – made the scene look almost bucolic, but her pulse was still racing.

'Are they watching other people at home, too? Or just you?' she asked.

'Everyone, I think. They raided the warehouse last night. Took the sound-system, and arrested Gaz.'

She didn't ask which Gaz because one Gaz was much like another. 'Jesus, Dan. Be careful, please. I can't have you going to jail right now.'

Dan took her hand and squeezed it. 'Relax,' he said. 'That's not going to happen. They're just trying to scare us.'

'Yeah. It's working.'

The squirrel had finished the Wotsits and was edging jerkily across the grass, hoping for a few crumbs of sandwich. Louis had fallen asleep. She wrapped him up and held him close, breathing in the unique comfort of his miniature warmth.

'Perhaps you should just . . . take some time out of the whole world-changing lark,' she suggested. 'You've done your bit now, more than enough. Let other people take over. We have other priorities.'

Slowly, with control, he put down his cup. His face was pale. His lips were tight. He rubbed his eyes. 'Okay, Sky, listen. You've got to understand something. I'm in this for the long haul. This is me. I'm not going to give up what I do just because we have a kid now.'

'That is not what I was saying. I'd never say that. I just think that perhaps you need a bit of time out. You're not enjoying it any more. Look at you: you're stressed all the time. We've got guys following us around when we're out with our baby. This is not okay. It has to stop.'

His head was in his hands. She held Louis closer, a tiny pulsing shield against her heart.

9.

'Do the kangaroo hop! Do the kangaroo hop! Do the kangaroo hop with me . . .'

She had come to pick up Dan and Louis from singing, so he couldn't wriggle out of the appointment again. He was, as always, the only dad there, sitting right in the middle of the semi-circle of mums, like a scruffy sultan with his harem. Louis was eight weeks old now, lying kicking on the mats in the middle. At the front was Anita, the sexy singing teacher, her hands poised in a hopping kangaroo pose. Dan's current favourite wind-up was swooning over Anita, who had dark hair and a lovely pouty smile and somehow managed to make even her kangaroo impression look alluring.

'Are you sure you don't want me to take him to singing this week?' she would ask every Monday morning.

'You have a break, love,' Dan would say, all generosity. 'I'll go.' Then he'd give her a wink, and mouth, *Anita*, in a loose-lipped, sensual manner.

On the other side of the door, Anita was launching into the final song. 'Bye-bye, Eddie, bye-bye, Jamila, bye-bye, Louis, it's time to go home . . .'

When it came to *bye-bye, Louis*, Dan sat him up and shook his tiny hand to make him wave. The door opened and one by one the mums wheeled their buggies out through the reception of the Hackney Downs SureStart centre, past the smart red bookshelves and stacked boxes of educational toys, and posters explaining now to apply for baby bonds. This was early New Labour Britain, when they were still building children's centres, still talking up this big new word *meritocracy*, a magical gleaming sorting system by which society would provide each and every one with exactly what they deserved. On what basis they deserved it, or didn't, well, that was unclear.

Dan was one of the last to emerge, pushing Louis in the navy blue buggy, having spent slightly too long bidding a fond goodbye to Anita. Skylark crouched to blow a raspberry on Louis's tummy, and when she looked up, Dan was frowning down at her with an expression that asked, What are you doing here?

'My God, you've forgotten again, haven't you?'

'Forgotten what?'

'You are unbelievable.'

She had been reminding him for weeks, but somehow it never seemed to sink in: they had to take Louis to the

Town Hall to register his birth and get the certificate. They were already way past the deadline, but every time she'd booked an appointment there had been some reason why he couldn't come: he was working, or he had a very important and absolutely unavoidable all-day planning meeting for May Day, or he was at the pub with Jez or one of the Gazzes and lost track of time.

'I'm really happy for you to do it,' he told her, every time she brought it up, even though she had explained at least ten times why that wasn't possible.

'You have to sign the birth certificate, Dan. Because we're not married. It's the law.'

'What I don't understand is why they even need to fucking know,' Dan would huff, although he knew it wound her up. 'It's, like, the tentacles of the state.'

She rolled her eyes, then stopped, because she was turning into Mother. Most people grow out of the *tentacles of the state* vibe, but Dan had grown into it. Most people come to accept with age that they are not entirely autonomous individuals but, rather, beings who exist in relation to one another in a structure of power, and that a certain amount of form-filling is inevitable, even beneficial. Even she, a formerly feral squat-bunny, had come to accept a form or two.

But Dan was now more militant, more idealistic, more paranoid than he had been when they had first met. He still didn't have a bank account, and although he lived with her, his name wasn't on the flat's lease. They didn't have a car. He had ranted at her when she got one of Tesco's new loyalty cards. *It's not about bargains, Sky, it's about the corporations tracking your every move. You're selling your soul for a cheap tin of beans.*

So, while his reluctance to register his fatherhood on an official town hall document was aggravating, it was not a surprise.

'The appointment's at three,' she told him firmly, grabbing the buggy. 'We've got to get a shimmy on'. Come on.'

He looked at his watch. 'Shit. I'm sorry, Sky. I have to shoot – I've got to meet Gaz at the police station.'

She actually stamped her foot. 'Dan! We've got an appointment! The birth certificate, remember?'

But he was patting his coat pockets, checking he had keys. 'Yeah, I know. I'm sorry. You just – I think you'd better just go ahead.'

'But—'

He was already walking away, banging through the doors of the centre and striding out of sight. She pushed the buggy into the toilets and stood in front of the mirror, gripped by fear and confusion, as her bones maintained their teeth-grinding hum, which was a little louder every day.

Louis opened his eyes and pierced her with dark blue baby wisdom.

'Okay. Well, it's me and you, Lou,' she whispered. 'Just the two of us.'

The registrar looked at her over the top of his glasses. 'And the father's name?'

'We're gonna leave that for now.'

The registrar was sitting on the other side of a grand mahogany desk, in a grand room, with peeling flock wallpaper. He had skin of a matching mahogany shade and his hair was grizzled white around the temples. He looked like a family man, a man who had six kids at home, and his name proudly on the birth certificate of each one. He was smiling kindly, but she felt sure there was pity in his eyes.

'I mean, I can tell you the father's name. Daniel Greene. He's my – we live together. I was with him just a minute ago.'

'Okay.' The registrar hesitated; his smile wavered. 'But the thing is, if you're not married we need him to be here in person. To vouch for his paternity, if you get me.'

'Yeah.' She coughed. She looked down at Louis's sleeping face. Another tear dripped onto his forehead. 'I get you.'

The registrar held out a box of tissues, and she took one. For a minute or two, neither of them said anything.

'I tell you what, my dear. We can leave it blank. And if you – if he – changes his mind, just make another appointment and we can fill it in.'

She nodded. Somehow, through a fog of tears, she signed the form.

As she left the town hall that day, bones humming, tectonic plates shuddering, two men with buzz-cut hair were sitting on a bench outside Moonlight Finest Kebab. Their double gaze followed her as she pushed the buggy towards the bus stop.

UCO122: So, an interesting development. I attended the black bloc meet-up last night, and managed to make some friends. I'm hoping to get precise timings and locations for May Day flashpoints. My new best friend Anarchist Ed tells me he has lovely like-minded individuals coming in from as far away as Germany. So—

DI Wells: Let me just stop you there, if I may.

[PAUSE]

As you will recall, your orders were to stay away from the black bloc meeting.

UCO122: But surely—

DI Wells: No, listen, Daniel. You defied my explicit instruction.

UCO122: Only because—

DI Wells: That's enough. Would you sit down? There is something we need to discuss.

[SILENCE]

This is going to be difficult, I know—

UCO122: Is this about those ridiculous goons you've had following me about?

DI Wells: It's about information that has been passed to me.

[PAUSE]

I'm aware that things have got out of hand. Very out of hand. If you get my drift.

[PAUSE]

It's in your own interests that this ends now. Before any further damage is done.

UCO122: [LAUGHS] Please, Martin. Let's not pretend that anybody here gives a flying fuck about my interests.

DI Wells: I can assure you that, as your supervising officer, your interests are always foremost in my mind.

[PAUSE]

Look. You know – and I know – that you have given your all to this job. That has been recognised right up the ranks. Yours has been one of the most successful tours this unit has ever had. Your reputation will survive this. But we need to draw a line under this . . . situation.

[SILENCE]

UCO122: Come on, now. You're not going to pull me out. You know it, I know it. Not yet. It's only a month until May Day. I'm in deeper than anyone, deeper than you'll ever get again. Look what happened in Seattle. You need me.

DI Wells: I don't need anybody who has divided loyalties, or who is compromising the security of every officer in this unit.

[PAUSE]

I think it's only fair to warn you that I'm going to discuss your situation with Bob and we'll make a decision about how to deal with it.

[SILENCE]

There's one more thing. You need to take a look at this, and fill it in by next week. It's come down from on high.

UCO122: [READS] "Welfare Survey for Undercover Officers".

DI Wells: It's a new initiative. Bob's been out in America, talking to the FBI about how to handle long-term deployments like yours. He came back with this. For a bit of feedback, you know.

UCO122: "Symptoms of stress for spies. Please tick the boxes to indicate whether you have noticed any of the following symptoms: development of a short fuse; uncharacteristic anger or resentment; significant increase in alcohol consumption." [LAUGHS. LAUGHS MORE].

Oh, God, Martin, this has to be a joke.

DI Wells: Not at all. I believe this illustrates my point about everybody here taking your interests very seriously. Please fill it in and return it by next week.

UCO122: Welfare! Tick the boxes! Oh, Martin. You couldn't make it up. Thank you for your concern.

10.

'No, leave it on,' he said, when she got up to switch off the light.

She stopped in her tracks, must have heard wrong. They hadn't even discussed the lights-off thing for years. It went without saying: it was always dark. As in not dim, not dusky or romantically low-lit, but blackout. She knew the shapes of his body with her fingers, the force and pressure of him. She knew how to let go, and how to close herself around him. But they never looked at each other, not like that.

But now, four years and one baby later, something had changed. *Leave it on.* She took her finger off the light switch, sat next to him on the bed.

'Are you sure?'

He nodded, and for a shocked moment, she thought he might cry. 'Yeah. Leave it.'

'What's brought this on?'

He didn't reply, just quickly, roughly, pulled her T-shirt

off over her head. She let him, lifting her arms, then sat to be looked at under the frank overhead light. Her breasts were heavy and full – Louis hadn't drunk much this evening – and the skin on her tummy was loose. But she didn't slouch or try to cover herself; she was what she was. She flicked the buckle of his jeans.

'Now you.'

He undid the belt and pulled off his vest. She undid his top button, holding his gaze. Traced a slow line with her mouth from his chest to his belly button, then went on down with her tongue. Nibbled at the elasticated waistband and tugged at it with her teeth. Every now and then glancing up to see how he was looking at her. It was a thirsty look, like he wanted to down her in one.

'Even better than I imagined,' she said, on inspection. Sat up, rocked him between her legs. He cupped her breasts in his hands, took a nipple in his mouth. He sucked gently and she felt her milk release. The other breast was wet now, too, so he licked it. He was completely absorbed with this sucking and licking, like Louis when he fed.

'Can we be gentle this time?' she asked.

He stopped and looked up, and once again the sad thing flickered. 'If that's what you want,' he said.

* * *

No doubt about it, something had changed, or was changing. He developed a new routine. When he was home, he would lie on the sofa for hours every evening, with Louis on his chest, curled up, like a snail. When he went away to work he would call her in his breaks, once a day, or even twice. His voice was always hushed on these calls, as though he didn't want his workmates to overhear the words.

What did Louis do today?

Did he nap?

And feed?

Has he turned over yet?

Give him a squeeze from me.

And then one day, when he was about to ring off, he said, 'You know, I love you, Sky. You and Lou.'

The words echoed across a vast space. He'd never said it before. He rang off before she could reply.

But if he was opening up, he was also shutting down. Black moods gripped him. He came back after being away for work, crashed through the door without even saying hello, stormed through the living room and into the kitchen, grabbed a can of Red Stripe from the fridge. He opened it and stood there necking it in gulps.

She was on the sofa feeding Louis, who jumped at the noise, and let out a wail. 'Hey! What's up with you?'

Ignoring her, he finished the can, crushed it in his fist, chucked it into the bin, opened the fridge again and took out another. Stomped past and out of the back door, threw himself into a chair on the balcony. She followed him, with Louis in her arms. The afternoon was fresh and springy; some of the tulips she had planted were coming up. Down below somebody was rattling a ball against the caged-in football pitch, *pang, pang, pang*.

'Shut the fuck up!' Dan yelled, and the noise stopped. He took another long swig, his legs jiggling.

She held Louis out to him. 'Here, have a baby. Breathe.'

Dan put the can on the floor and took his son in his arms. The soothing effect was instant. He inhaled Louis's sweet baby smell, closed his eyes. A tear ran from the corner of his eye onto his son's head. She crouched beside the chair and touched his hand. 'Dan, please tell me, what's going on? Is it the police?'

She hadn't seen them for a few days, now. Every time she left the flat she checked, looking up and down the road for their twin gazes, their matching buzz-cuts, their navy blue car. When they were there, it was obvious: they didn't even try to hide. Dan said they wanted him to know he was being watched.

He didn't open his eyes, just held Louis and breathed.

She put her arms around both of them. She was not used to being the one holding it together. After a while, he looked up and shook his head, as though he was shaking something away.

'It's okay to be scared, you know.'

He didn't reply.

'Remember, you can stop this. You don't have to be involved. Just tell the others you need a break, you can't do May Day. They'll understand.'

He sniffed, wiped his nose. 'If only.' He let out a long breath. 'It's too late, Sky. Decisions got made, a long time ago.'

'Which decisions?' she said. 'By who?'

But this only made him cry more, proper sobs now, shaking his broad, muscled shoulders.

They sat there in silence, her hand on his. Louis snuffled against his neck, and he leaned his head on his son's, eyes closed, as though he could absorb him through his skin.

Eventually he said, 'I might have to go away for a while.'

The football started up again, *pang pang*, but now neither of them cared. It was a relief to have some, any, distraction.

'Things are too hot at the moment,' he said. 'It's not good – not for me, not for you. Or for – or for Lou.'

'What do you –' she said '– but you can't –' and finally – 'how long for?'

He didn't reply.

'You're coming back, right?'

Still, no answer.

'But you said you wouldn't leave.'

Pang pang. Pang pang. There was a world out there, in which other families were playing football, cooking their dinners, watching *EastEnders*. She'd never wanted to be like them. She hadn't wanted Tupperware, never cared about parking: she'd wanted love and pain, and she had found them.

They sat there for a while longer before she said, 'And your family?'

'My family?'

'Have you told Diane and Sidney about Louis? Perhaps they could offer us, you know, some support.'

Now through the tears he burst out laughing. A dry, barked laugh, with a sob somewhere in it.

'Don't you think your dad would want to meet his grandson? Don't you think Louis needs to meet him? Especially if you – if you—'

She didn't say *if you're not here.* She couldn't.

'Believe me,' he said, his voice strenuously level, 'I wish that was possible.'

'But it is! Just do it, Dan. It's time to let us in. Take the wall down. We're your family now.'

She was trembling. She reached over and took the

baby from him, hugged him to her chest. At the sudden movement Louis's face went purple with outrage, and he screamed furiously. 'You can't go on like this,' she said.

There was a pause before Dan sighed and grasped his forehead in his hand, pinching his temples hard. He said, with such sadness in his voice that she instantly regretted her anger, 'You're right, love. I can't.'

11.

THERE were moments, during those first months of Louis's life, when the sun peeped through, moments when Dan would be seized by some memory of affection, and change a nappy, give her a hug, crack a joke, play and laugh with his son. In the morning, before the day had really set in, he could be almost normal. They'd lie in bed with Louis and watch him perform his latest trick: he loved to lie on his back and kick his legs, one after the other, with this serious expression, like a sportsman doing his morning exercises.

Dan had a knack for making Louis laugh: he'd nuzzle his neck with a stubbly chin, blowing raspberries, and Louis would grab at his nose and shriek with joy.

But then he'd go cold. One day, she left Louis with him while she popped out to get nappies. On the way back she could hear cries as she turned into the entrance of the estate. Inside she found Dan lying on the sitting-room floor, blank-faced, unmoving, while Louis howled from the bedroom. She rushed in to find her baby red-faced and hiccuping in his cot. As she patted and soothed him, his tiny body rigid with fear, Dan remained staring at the ceiling.

After that, she wouldn't leave them alone together.

At Louis's development check, she tried to bring it up with the health visitor. She was a new one, just another face on the parade of different healthcare professionals, a large, stern, flush-faced lady in a flowered blouse, whose name badge read 'Jean' and whose ample backside spread over the sides of her wheelie chair. Jean spent a few minutes testing Louis's hearing and responses, ticking boxes in his little red book. Then she sighed, picked up a clipboard from her desk. 'Right, Mum. I have a few questions about your own health.'

'Okay,' she said, assuming Jean was about to give her some pelvic-floor exercises, or tell her to drink full fat milk.

'What has your mood been like since Baby arrived?' she intoned, in a brisk monotone, her eyes fixed on the clipboard. 'On a scale of one to ten, with ten as "very happy".'

She thought back over the months since Louis was born: the animal exhilaration of birth, the pride every time he smiled, whenever he did something new, the nagging fear of Dan's mental state, the intermittent bone-hum, the exhaustion and worry she seemed to be dragging around wherever she went. She grasped in vain for a number to represent it all.

'Perhaps – a five?' she said hesitantly. The health visitor nodded, scribbled on the clipboard. It occurred to her that five was too low, that the health visitor would realise things weren't right. 'No, sorry, a nine.'

Jean pushed her glasses down her nose and looked at her over the top. 'Which one? There's quite a big difference between five and nine.'

'Oh. Nine,' she said. 'Definitely nine.'

Jean looked back to her clipboard. 'Any depressive feelings?'

'What, me?' she said, startled.

'You.' She avoided Jean's gaze, looking around the room with its blue lino floor, its peeling green walls. 'Any suicidal thoughts, plans, ideation? We ask all new mums the same questions at twelve weeks. So we can offer support with post-natal depression.'

'I don't think so. But –' Louis woke up and wriggled in his sling; she took him out carefully and put him to her breast '– what about fathers?'

Jean frowned. 'Fathers?'

'Do they ever suffer from post-natal depression? Is it

possible, for example, for the birth of a child to reawaken traumatic feelings and memories that a father has from his own childhood? If he were, say, abused as a child, might he be so worried about the same thing happening to his own child that he would kind of switch off his feelings? To sort of defend himself?'

Jean glanced at her watch: they were overrunning the ten-minute appointment. 'Well,' she said, 'I don't know anything about that.'

She unclipped the checklist, scribbled something on the top, folded it and put it in the paper tray on her desk. Then she looked up, apparently surprised to find that they hadn't gone yet.

'Was there anything else?' she asked.

She put Louis onto her shoulder, inhaling him, hoping to infuse her brain with milk, talcum powder and daisies. 'No,' she said. 'That's all.'

12.

I T wasn't just Dan they were watching. She realised that one afternoon in Ridley Road market. The air smelt like smoggy rain, with top notes of fish and rotten fruit. Reggae thumped from one of the dilapidated shop fronts, and a flag fluttered outside, red, yellow and green. One side of the cobbled street was a little Jamaica, stalls laden with string vests and tams, plantain and okra, goat carcasses and – could this be legal? – a wicker basket full of live giant snails. On the other side, white traders had staked out their territory with piles of fruit and veg, rubbing their hands and calling out, 'PAAAAAAAAND a bowl,' never breaking a smile.

And that was when she saw them, buzz-cuts and dark T-shirts, standing on the white side, one idly fingering a bowl of pomegranates, the other looking about with darting eyes. Their matching hair and broad shoulders, their neat jeans. The darting-eyes one saw her and nudged the pomegranate one.

Her heart beating faster, she walked on.

She wasn't even with Dan, and they were here. Why? She was nobody, any more. She didn't do anything, except feeding, wiping, cooking, shopping. What possible interest would she be to anyone? Why would anyone be watching her?

Perhaps the health visitor had seen something she hadn't realised: she must be depressed, she must be paranoid. She reached the corner and glanced back the way she had come: the flag was still fluttering, the reggae still bumping, the market still bustling, but the two buzz-cuts had melted away like ice.

DI Wells: Right, Daniel. Sit down, please.

[PAUSE]

I have spoken to High Command about the
concerns we discussed last time. And, after
thorough consideration, we have decided, on
balance, that we have no choice other than to
end your tour.

[SILENCE]

I'm sorry, Daniel.

UCO122:   When?

DI Wells: Soon. As in, start your preparations for with-
drawal now.

UCO122:   But I—

DI Wells: Those are your orders.

UCO122:   Fuck this. Fuck orders.

DI Wells: Okay. [LOWERS VOICE] I'm going to speak
frankly, Daniel. This is the only possible course
of action. Do you hear?

[PAUSE]

Any other option is a worse option. Worse for
you, worse for us. And actually worse for –
them.

[PAUSE]

> They'll be okay. Listen to me. We'll make sure that they are.

UCO122: How? By watching them? By sending your stupid brainless goons to follow them around?

[PAUSE. WHISPERS]

> I can't leave.

[SILENCE]

DI Wells: Daniel, I realise this is going to be difficult. But look at it from their perspective. You can hardly live as you are for ever, and there's no way you can tell them the truth without compromising this entire operation and under-cover officers all over the country. You know that as well as I do.

[PAUSE]

> Anyway. For them, for your girlfriend and your baby, the truth would only cause endless pain and confusion. The best thing is for them to move on with their lives, without ever knowing what has happened here. Without knowing who you are.

UCO122: They know who I am.

DI Wells: What? Have you told her?

UCO122: Of course not, Martin. I didn't mean it like that.

DI Wells: We can't let this go on, Daniel. It's not an option.

[PAUSE. LOWERS VOICE]

The longer you leave it, the worse it will be. Get out now, and don't look back.

[SILENCE]

UCO122: How long do I have?

DI Wells: We can't pull you out before May Day. It would create too much suspicion. But as soon as possible after that.

UCO122: And after May Day we can take stock.

DI Wells: No. Look at me. No, not at the floor, not at the wall, really look at me, in the eyes. Right. Okay. This is happening.

UCO122: I'm gonna talk to Bob.

DI Wells: Believe me, you don't want to do that.

UCO122: But—

DI Wells: But nothing. I have given you an order. Understood?

[PAUSE]

Daniel?

[SILENCE]

13.

Oh, happy (May) Day! Her dearest darlingest Revvy, her long-lost friend, was back in town for the now-annual world-changing carnival. Sitting in the living room with Louis propped awkwardly on his black-denim thighs, he looked as he'd always looked: strange. Skinny as ever, but with a leathery sheen to his long cheeks, evidence of nearly a year living on the road. His clothes had taken on more colour, an air-force coat with the collar turned up, and long cherry-red DMs.

'Are you back for ever?' she asked, after hugging him extremely tightly. But he wasn't.

'Just for a visit. I wanted to support your man in his ambitions. But I'll be off again, my dear.'

'Off to where?'

'Belgrade,' he said. 'For now. Then who knows?'

Seeing her old friend again was like having a hot meal after too long living on crisps. He was the same old Rev, equal parts ridiculous and imperious. Louis reached for his nose, and he took the tiny hand in his mouth and pretended to eat it.

'Did I tell you the art man from Glastonbury called *me* this year?' he said, when he'd finished nibbling. 'Check this: they want us to take over a whole area next to the Green Fields. What do you think I should do with it?'

'Fill it with toilets,' she suggested. 'They never have enough.'

'That's not a bad idea.' He put his arm back around her, gave her a squeeze. 'So what are we thinking? Shall we head out to this shebang today? Dan's already there, right?'

'Right.'

He had left early that morning in one of his dark moods. She'd said, 'You don't have to go, you know, just stay,' and he had turned his mouth down and said, 'Let's not talk.'

'I hear this one is all about laughter yoga,' Rev was saying. 'Getting back to basics. About time, too. It's all been far too serious.'

'Actually,' she told him, 'I had thought about giving it a miss.'

Rev shook her gently. 'Come on, Skylark McCoy. What happened to my warrior queen?'

'Not sure,' she admitted. 'I haven't seen her for a while.'

'Nonsense,' said Rev, pulling his arm away. His pale eyes stared piercingly into her tired blue-green ones. 'She's in there, I can hear her. Listen.' He adopted a high-pitched pleading-captive-warrior-queen voice: 'Let me out, let me out.'

Louis was eyeing him suspiciously – perhaps he wasn't used to adults having less hair than him. 'We can take this little man and hang around on the fringes,' he said. 'Don't worry, we'll get out of the way as soon as it kicks off. Which it obviously will. Laughter yoga, my arse.'

She laughed, and nestled back into the crook of his bony arm. 'All right, then. Just for you.'

Louis looked from him to her, and back again. Perhaps she was imagining it, but he looked more relaxed with Rev around. Lately he'd developed a furrow between his brows, a look of permanent puzzlement that seemed wrong on his three-month-old face. But now he was smiling pudgily, correctly, like a baby in an advert. That smile loosened some knot in her stomach that she hadn't known was there.

The weather was doomy: grey and heavy. Police lined the exit to Westminster tube, and in Parliament Square the low chuck-chuck of helicopters nagged like a head-ache. But there was a cheerful and determined crowd on the green. People were dressed as butterflies, and in pagan costumes, leaf masks and green capes. There was a woman on stilts wearing a unicorn hat, and on a plinth by the gates to Parliament someone had erected, for no obvious reason, a giant penis carved from a large tree trunk.

'Ah, the Great British Woodcock,' said Rev, leaning in to caress its straw pubic hair fondly. 'An endangered species, I believe.'

Through the crowd, they glimpsed a familiar shimmer of red hair.

'Eefy!' called Rev. Aoife looked around, then dropped the spade she was holding and came charging towards them, nearly knocking Rev over with a leaping straddle. 'I thought you were off with the circus!'

'I've run away to the real world. Just for the day.' She grinned, kissing him and setting herself down. 'I wanted to be here, you know, to support. Let bygones be . . .'

'Exactly,' said Rev. 'I mean, all power to them. Keeping the fire burning.'

Aoife turned to Skylark, peeked in to look at Louis, asleep in the sling. 'My goodness. Look what you made, Mama Sky,' she said, giving her a tightness of hug that's only possible if you've spent years training for the flying trapeze. 'Aren't you clever? He looks just like Daddy, doesn't he?'

He did. Louis had Dan's dark colouring and his wide mouth. Aoife stroked his sleeping head.

The plan Dan and the Jez-Gazzes had come up with was to turn Parliament Square green into a market garden. Aoife led them over to a group of people with spades, who were digging up a large, muddy square of turf. A wheelbarrow appeared filled with seedlings: rhubarb, chard, herbs. Carefully they took turns to dig holes for the plants, while Aoife unhooked a megaphone from her golden goddess-belt.

'Once upon a time, all the land this city is built on would have been held in common!' she announced in her singsong voice to the crowd of spectators, while a woman draped in flower-fairy chiffon danced around the mud patch. 'It was used for the good of all: to grow food, to graze animals, as space for children to play. Today, this common land has been sold off to the government's crony corporations. They build on it, tarmac it, rent it back to us at an extortionate price. Today we are acting as witnesses, showing that we know the truth: whatever their expensive lawyers and their precious bits of paper say, this land can be bought and sold by no man!'

The crowd cheered, and in the melee a guy with one long matted dreadlock removed his jacket, then stripped off his tie-dye leggings and string vest. Naked as a mole rat, he shouted, 'Up the Diggers!' and threw himself face-down into the muddy patch, thrashing about until his bony white body was fully coated.

'He's squashing all the plants,' Aoife muttered to Rev.

He tutted. 'Fucking hippies.'

The crowd around them pressed forward, everyone straining to see the naked guy. He stood up, grabbed a strip of turf, and sprinted towards the nearby statue of Winston Churchill. With muddy buttocks gleaming in the spring sunshine, he scaled the plinth and managed to hoist the turf onto the statue's head. The nation's revered wartime leader scowled from beneath a jaunty green mohawk.

A gaggle of TV cameras had materialised; from somewhere in the crowd a bongo started up.

'That haircut is too good for him,' remarked Rev.

Louis needed a feed, so they broke away from the protest and strolled down Whitehall, looking for a quiet spot. Along one of the side-streets an old-fashioned greasy spoon was open, tables outside covered with checked plastic cloths. They could sit pleasantly drinking tea, watching the protesters pass by at the bottom of the street: a guy with glasses holding a banner reading 'Global Capitalism Isn't Very Nice', two women in pink and silver unitards, a man blasting out techno from a sound-system mounted on a bicycle.

Rev took out his smoking apparatus and crossed his long legs. He leaned back and looked at her in a long, unbroken silence, the kind of silence that erodes a person's defences. 'So,' he said at last. 'How's Dan?'

And there it was, the question she had been equal parts dreading and hoping for. She hugged Louis and took a careful sip of her tea.

'I think he's losing the plot,' she said simply. Her throat tightened, with a feeling like nausea, but words came spewing out instead. She told Rev about the driving licence, and the story about the false name. She told him about the town hall, and the buzz-cut men. She described the bone-hum, the tectonic plates. She told him that Dan had been talking about leaving.

'I don't understand what's happening,' she said. 'Nothing makes any sense. He says he loves us, and I think that's true, but then in the same breath he says he's going somewhere and never coming back. Every one of his explanations just confuses me more.'

Rev stubbed out his joint, and started to make another. 'What's your instinct?'

She closed her eyes, to connect to that deeper part of herself, the part she had been repressing and ignoring, telling it to shut up and not mess up her life. Was it still there, the inner landscape, the ancient trees and standing stones?

'He's hiding something,' she said, and as soon as she did, some heavy thing lifted slightly, some clouds parted and a tiny sliver of light shone through. 'Always has been. And not just a small thing. Something fundamental.'

'Well,' said Rev, 'you must trust yourself.'

There was a shout from the end of the road, where the crowd had grown denser. A row of yellow-jacketed police stood across the end of the street. She and Rev watched the scene from just a few metres away, like spectators at a seaside show. There was a crash, the unmistakable smash of a plate-glass window, followed by a chorus of shouts. The crowd on Whitehall started to run and scatter.

'Oh, boy, there goes McDonald's,' said Rev, standing up and grabbing her nappy bag. He turned to give her a don't-panic smile. 'Time to leave.'

Getting back to Westminster would have involved negotiating a scrum, so they walked fast along backstreets, towards Charing Cross. The helicopters were low now, slicing at the air above their heads, drowning the sound-systems. As they crossed the Strand, a group of men pushed past them, dressed in black hoodies, their faces covered with balaclavas and bandanas. As they neared the station, a grim line of riot police filed past, shields lifted.

'Ah, the Angries, Lord love them!' said Rev. 'They must be coming for the laughter-yoga workshop.'

Back at the flat, they watched the TV-news version. There was a brief shot of Winston's mohawk, and then some shaky footage of the McDonald's on Whitehall. Three balaclava-bandana men were hurling chairs at the window, but you could hardly see them for the photographers.

'How come they left McDonald's open?' she said. 'After what happened last time?'

'Bait,' said Rev. 'And those bell-ends went straight for it.'

The prime minister's teeth appeared on the screen. *These actions have nothing to do with belief or conviction*, said the teeth and their surrounding face, *and everything to do with mindless thuggery.*

She switched it off in disgust. 'Well, that's that, then.'

'Fucked?'

'I think so. Don't you?'

'Yep. For now, at least.'

There didn't seem to be much else to say. It wasn't worth remembering, right now, how much the world-changing group had once meant to them. Those times when it had seemed they were approaching a moment, after which they would no longer be the weirdos, that in fact the weirdos would be the ones who saw no problem with the way things were.

Instead they watched Louis, who was lying tummy-down on a padded floor mat she had found in a charity shop. He reached out to grasp a yellow stuffed caterpillar, missed, and tried again. He shuffled forwards by rocking his body, just a tiny bit, reached again, almost tipping himself over.

'This is a long game,' said Rev. 'We'll get there. Eventually.'

Silence stretched between them, broken only by Louis's snuffles and grunts, before Rev went on, 'You know what I think? What we were doing, back then, was planting seeds in people's minds. And it's too early to say when and if and how those seeds will grow. Perhaps there was some kid there today, and she'll be the next one to take it on. Perhaps she'll get us over the line.' He nodded at Louis. 'Or maybe he will, when he's all grown-up.'

Louis reached for the caterpillar, and missed again.

UCO122: Why was that McDonald's left unguarded?

DI Wells: Sorry?

UCO122: In all my reports I made it clear that McDonald's would be a target. My explicit recommendation was that effective policing would involve closing – or at the very least guarding – each local branch of McDonald's. And yet there it was: shutters up, not an officer in sight. Was it deliberate?

DI Wells: As you know, Daniel, I'm not directly involved with strategic planning around the policing of demonstrations. I wouldn't know why.

UCO122: Let me answer that question for you: of course it fucking was.

DI Wells: Your conjecture is as good as mine here. But as you can imagine, with the sensitivities around the new Terrorism Act, it may have been judged expedient to—

UCO122: To leave that particular target wide open, so that the black bloc could trash it, giving the government all the reason they need to criminalise political protest.

DI Wells: I don't know.

UCO122: I think we both know, Martin.

[SILENCE]

> I mean – what have I actually been doing, for all these years, in the field? I've been telling myself that the information I provide helps to make sure that policing is appropriate. That nobody gets needlessly beaten up, or tear-gassed. Blessed be the peacemakers.

DI Wells: And you have done an excellent job.

UCO122: But that's nonsense, isn't it? Because it's all twisted. We needed May Day to get violent again in order to justify our own existence, in order to support the government in passing more repressive laws.

DI Wells: Now, Daniel, that is not—

UCO122: Come on! You know it as well as I do. This isn't policing – it's politics.

DI Wells: Inevitably, our work has a political dimension.

UCO122: And the politics is completely focused on protecting the right of an elite few to run the economy as they wish, while everyone else . . .

DI Wells: [LAUGHS] Now you really are sounding like one of them.

[PAUSE]

UCO122: I don't know what I am any more.

DI Wells: Daniel, listen. I understand that this is difficult for you. You just have to hold on to the fact that soon you will be back in your normal life. Things will settle down. You'll see all this from a different perspective.

UCO122: No, Martin. Don't you see? I left normality behind for ever when I took this job. Believing I even know what normal is – that's a luxury I don't have any more.

14.

THE morning after May Day, it was 6.30 a.m. when she heard the door, then saw Dan standing at the end of the bed. His eyes stood out in his pale face, and he was licking his lips.

'All right, love,' she said sleepily. 'How was it?'

'It was okay after I took three pills.'

'That good, eh?' He sat down on the end of the bed and put his head into his hands. 'We were there for a bit, and then I saw the rest on the news.'

She lifted the corner of the duvet and he crawled in beside her with all his clothes on. He smelt unlike himself, smoke and something sharp, alcohol or old sweat. She sighed and turned over and he draped his strong clammy arms over her. 'We're supposed to be going to Henfield today. Mum invited us for lunch.'

'Tell her I'm sorry.'

'I'll tell her you're off your head on drugs.'

'Whatever. I don't care. Tell her anything. Sky . . .' He

turned to face her. His eyes were red and heavy, the pupils huge. He whispered, 'Forgive me.'

She soothed him, shushed him sleepily. 'Stop it, Dan. Let's talk about it later. You're just buzzing.'

Mother's knife scraped against her plate.

'Was he one of those men on the news,' she said, 'throwing the chairs?'

'Oh, Mother. Don't start all that again.'

'I'm not starting anything,' said Mother, and took another mouthful of roast potato. 'I was hoping he'd be here,' she said. 'There's something up with my computer.'

'I could have a look at it, if you like.'

'Oh, no,' she said. 'I'll ask Dan. Next time.'

On the sideboard, the carriage clock ticked loudly, and either side of it the baby pictures stared out, one plumpy, one serious. With every tick an iron band tightened around Skylark's head. She had thought it might be restful to spend the day in Henfield – since having Louis she'd found herself oddly drawn to it, to its quiet long roads, endless afternoons, well-trimmed lawns. But she felt tense today, as if she was waiting for something. She looked at the plate of roast meat, gravy and potatoes in front of her: unappetising, heavy, greasy.

'Anyhow,' said Mother, 'I've put some kale in. And some sprouts, for next Christmas. Now, Lilian, I've forgotten, do you like beetroot?'

Skylark opened her mouth to say something about beetroot, and closed it again. She didn't want to talk vegetables any more. 'Mother,' she heard herself say, 'did you ever think Father was going to leave you?'

Mother looked at her beadily, chewing. When she had swallowed she put her knife and fork down neatly, side by side. 'He did leave me,' she said. 'For three months. After, er, the . . . Before you were born.'

'Really? I had no idea.'

She took a large slug from her wine glass. 'He left me after your big sister died. After Janie died.'

Tick-tock, tick-tock. Two babies in their frames. On the right Janie, the big-sister-who-never-was. The perfect pinky plumpy baby, who, had she lived, would have loved Tupperware and developed strong views on parking. Who would have cared about dahlias and stayed in Henfield and visited Mother every single Sunday. On the right, an awkward, skinny, serious baby: a baby who already feels the weight of the world, and who will run away, to try to change it.

'He went off around America on a Greyhound bus. I had no idea if he was coming back. But you see . . .' Mother waved her hand as if she was swatting a fly

'. . . it was different for us,' she said. 'We were brought up knowing that men needed their freedom.'

Skylark looked down at her plate. 'And do you still think that?'

Mother frowned, concerned. 'Why, Lilian? Are you having trouble?'

'Dan's been talking about going away. I'm not sure how long for.'

Mother took her daughter's hand, her grip tight, her mouth set in a determined line. 'You must let him,' she said. 'That's all you can do. If you can let him go, he may come back. Like your father did. He went off to America, got it out of his system, and then he came home. He came home, because . . .' she smiled, relaxed her grip '. . . he knew that he wanted you.'

15.

S HE knew that something had changed in the flat as soon as she opened the door. The hall looked lighter. She bumped the buggy over the doorstep and realised that all the photos were gone. Next to the Tofu Love Frogs poster, the wall was bare, speckled with grey spots where the sticky-tack had been.

Then she noticed the corkboard from the kitchen, propped up on the floor. It was stripped bare: every single flyer, postcard, old birthday card and shopping list had been removed. There was a neat heap of drawing pins on the shelf above the radiator.

Oh, God. She knew then, she knew.

The pile of shoes by the door was smaller, and only her tattered old orange puffa hung on the pegs.

She left Louis sleeping in the hall and walked steadily, one foot in front of the other, into the bedroom. She opened the man-cupboard. Sure enough, all the junk, every scrap and screwdriver, had also gone, and on the top shelf, there was a space where the yellow box used to be. His drawers: empty. His bedside table: cleared. No dumbbells on the floor; no drill bits or baccy pouches or broken carabinas. He'd taken all the junk, every single bit of it.

She sleepwalked through the sitting room and into the kitchen, where a note lay on the table.

Sky, I'm going to post you a letter.
You'll get it soon. I love you, both of you. Believe it or not.
D

'Fucker,' she said, 'fucking goddamn fucker.' Turned her head to look slowly, dazedly, around the kitchen. The photos on the fridge: gone. The cookbook that he had scribbled in: gone. Her address book was lying next to the note on the kitchen table, and she picked it up. There was a page at the front where she had always written down Dan's contact numbers at work, when he went away.

It wasn't there. He had torn out the page.

She put it down, the thumping silence in the flat pressing in around her. Hands on her face, just to try to contain the rush, to try to protect herself from the message that her home was delivering, from every wall and bookshelf and cupboard, unmistakably: Dan hadn't gone off just-for-a-while. He wasn't going to travel around America on a Greyhound bus so he could come back in three months, ready to settle down.

Daniel Greene hadn't just left: he had disappeared.

IV.

2010–11

9. AFTERCARE

...

9.2 RETURN TO LIFE AFTER AN SDS TOUR

9.2.1. First of all, ask yourself the following questions:

Q. Why does my suit not fit?

A. Because you are fat.

Q. Why do I have to get up at 7.30 a.m. every day?

A. Because they will stop paying you if you don't.

Q. Why do I have to get off the tube with the rest of the lemmings?

A. Because they took your van off you.

Q. Why am I poor?

A. Because you've got used to spending dosh you no longer have.

> **Special Demonstration Squad,**
> *Tradecraft Manual*

1.

Louis took a comedy-huge gasp of air, and blew. He got every candle out and then collapsed in an am-dram heap on the table, gasping. The small huddle of people around him broke into applause, then laughter when, one by one, all ten candles fizzed and sparked back into life. He raised his eyebrows at the woman whose springy hair now had one streak of grey at the front.

'Never gets old, does it, Mum?'

She grinned and ruffled his dark curls. 'Unlike you. Double figures, hey. Would you just stop growing up so fast?' She offered him the knife. 'Go on, then, make a wish.'

Louis put the knife on top of the cake and paused for a second.

'Hmm, let me read his mind,' said Rev, squeezing his eyes closed, fingers to his temples. 'He's wishing for . . . is that Harry Potter?' His eyes snapped open. 'Come on, now, Louis, you know we couldn't get you a wand from Ollivanders – we're just a bunch of Muggles.'

'That's not it.' Louis reached over and bopped Rev lightly on the top of his shiny bald head. 'I know what I want to wish for.' He pushed down on the knife, and for one delicate moment, the room went completely quiet. A soft clink as blade met plate. He looked around, with the funny serious expression he often had, the one that looked too old for his face.

'Done.'

They passed the cake around – it was chocolate brownie with chocolate icing, Mum's secret recipe, honed over a decade of birthday parties. Suze topped up the prosecco – everyone was drinking out of chipped, tea-stained mugs, because Mum had forgotten the glasses, then had to root around for drinking vessels in the play scheme's staff kitchen. Granny offered around the allotment raspberries she had put in the freezer last autumn; they were Louis's favourite fruit, and she always saved some for his birthday.

Outside it was grey and not that cold for winter. Louis and Franklin raced around the playground as the little group inside hoovered up the cake, licking the last smudges of chocolate off their fingers. This was a tradition: they always celebrated Louis's birthday in the play scheme, with the same people: Mum, Uncle Rev, Granny, Suze, Louis's best mate Franklin, and Franklin's mum, Aoife. On his first birthday, Suze had let them use the place for free, as a special present to cheer Mum up. But they'd stuck with it ever since: the flat was too small for parties and they were too broke to hire a hall.

Secretly, Louis had hoped for something different this year. The play scheme was too babyish for a ten-year-old – he would have liked a party at LaserZone or Lee Valley Ice Rink, with some of his friends from school. He wasn't a little kid any more: he didn't want to play in the padded, damp-smelling soft-play room, the ball pit, or in the playground with its once-magnificent wooden pirate ship. But he didn't say that to Mum, because she would be upset. *We do our best for you, Lou*, she said, whenever he let slip that he wanted something she couldn't afford, like a trip to the Harry Potter studios in Watford.

But that wasn't what he had wished for.

When Franklin went inside for more cake, Louis sat down on his own under the tall plane tree. As he watched the leaves move, a shifting puzzle made from light and shade, he scratched absently at the itchy place on his arms, the inside of his elbows. The itch was getting worse: whatever he was doing now, whoever he was with, it was always there. Sometimes it was so bad that he scratched until his arms bled, and Franklin teased him, 'You got fleas, Lou?'

When it started Mum took him to the doctor, to see

if it was eczema. The doctor looked a bit puzzled, said the skin underneath the scratching was fine. He gave them some cream, and Mum put it on every morning and night.

But the itch didn't go away.

It was getting worse.

Louis sat there beneath the tree on his tenth birthday, scratching and scratching and thinking about his wish.

'Big man.'

He squinted up to see her towering over him, her mass of springy hair blocking out the light. Her cheeks were flushed, and she was swaying ever-so-slightly, which was something Mum did when she had drunk wine. She extended her arms and balanced unsteadily on one leg, in what she liked to call 'the crane pose'. Mum knew nothing about martial arts: she'd just watched *Karate Kid* once.

'Let's rumble.'

'Oh, God,' sighed Louis. 'Not again.'

'Damn right. I'm going to take you down to Chinatown.'

Louis hopped to his feet, and Mum dragged him inside, past the office, the book corner and the craft table, through the door into the soft-play room. The walls were padded and covered with a shiny PVC material, yellow, green, red and blue. The floor was lined with crash mats. There were a few battered foam cushions strewn around, triangles and arches, probably the same ones Louis used to climb on when he was a toddler.

'Ten Schmen, I can still have you,' Mum growled, putting her hands up. She did a high kick – not bad for an old person – but Louis easily caught her foot in his hand. He did judo at school, so he actually knew what he was doing. He held Mum's foot there, taunting, for a moment or two, before flipping her onto her back and pinning her down on the crash mat.

'You see,' he said, raising his hands in mock-apology, 'it's too easy.'

But Mum surprised him: she wriggled out, sprang to her feet and charged, drunk and reckless. Louis was caught off-balance. He staggered and fell, and she threw herself down next to him. They lay side by side for a moment, panting and staring up at their reflections in the mirrored ceiling. They had features in common, mother and son: the same compact, energetic body, the same curls.

But Mum's hair was blonde with a grey streak that would make most people look old but she managed to make look almost punky, because that was her personality. Louis's was darker, and their faces were different: hers heart-shaped; his squarer, with a gap between his front teeth and a dimple on his chin. *You look like him, Lou*, she told him sometimes, and her voice was always so sad when she said it that he wished she wouldn't.

Louis lifted his hand to start scratching, then stopped himself and took a deep breath. 'Hey, Mum, I wanted to ask you something.'

In the overhead mirror, she frowned. 'Don't. Let me guess. You've got a crush on a girl and you want my advice. Now, let me see . . .'

Louis's face contorted in outrage. 'Ugh! Of course not!'

'Sorry! Sorry. It was the first thing that came into my head. You looked so grown-up all of a sudden.' Mum sat up, fiddling around, trying to tame her hair into a hairband. 'What is it, then?'

'I wanted to ask you about Dad.'

She froze briefly, then dropped her hands to her sides. 'Okay.' She looked down. 'Of course. You know you can ask me anything you want.'

Louis swallowed, and started to scratch. 'I want to find him.'

'Oh, Lou.' Mum hugged him. 'I have tried. You know that, don't you?'

'Yeah, you tried ages ago. But it's different now, isn't it? With social media, with Facebook, and all that. It's not like it was back then. No one can just disappear any more.'

Mum sighed. She gave up on the hairband, pinged it

across the room. 'The thing is, love, that Dad could have found us any time he wanted to. He knows exactly where we are – we're still living in the same flat. I know it's hard to think he might choose not to.'

Louis turned towards her, and she saw something new in his face – a look of determination that, she realised, with some satisfaction, definitely came from her. 'But I don't have to leave it up to him,' he said. 'I get to choose, too.'

There was a silence while she considered her son: he thought he was so grown-up, but he was still a baby, really, with no idea of how the world worked. He still believed in magic, for God's sake.

'I just don't want him to disappoint you, love.'

'I don't want *me* to disappoint me,' he replied, and there was really no way to respond to that, except to hug him really tightly.

2.

'Louis wants to find Dan,' she told Mother, once the others had all given them tipsy, nostalgic hugs and left, and they were doing the last bits of tidying up. Louis had gone home with Aoife and Franklin, to watch a film.

'Of course he does,' Mother said. She had brought her own pinny, a Laura Ashley print number, and was up to her elbows in soapy water, washing cups and chocolate and raspberry-smeared plates, stacking them in sparkling orderly heaps for her daughter to dry. 'He's growing up. It's natural that the boy wants to know his father.'

Skylark stopped drying. She stared down at the floor, picking at the damp tea-towel. 'But I'm really scared,' she said. 'Of what we might find.'

Mother sighed and clanked another cup down on the draining board. 'A dose of reality might be healthy for him,' she said. 'Better than believing his father is a perfect human, some great revolutionary hero.'

She said the last three words as though they tasted bad. 'I never told him that.'

Mother pursed her lips. 'Maybe not in so many words, Lilian,' she said. 'But you've allowed him to believe it.'

Skylark sighed and sat down on one of the strange uncomfortable spaceship-shaped chairs. Ten years' experience had taught her that it was most comfortable to lean forward. Which was also the best position for when she thought about Louis, because considering her son in the abstract caused a physical sensation of pain around her stomach, her liver and her lungs. It was the ache of love and guilt so closely entwined that they were in fact one and the same thing.

It was true that she had encouraged Louis to think the best of his father. She had always spoken highly of Dan, taken care not to run him down. *Daddy was fighting for a better world,* she had told him. *And that's not an easy thing to do. There are lots of very important people out there who want things to stay as they are. He made some of those people very scared and angry, and they wanted him to go to jail. That's why he had to leave us.*

Louis loved this story, like a fairy-tale or a Disney movie, *Tell me again, Mum,* and he'd listen with his big dark eyes open wide.

He was doing it for you, Lou, for your future. Because he loved you so much.

She wanted him to think the best of his father so he would think the best of himself. And it had worked. Louis was a serious, thoughtful child, a child who saw himself on the side of the good. The kind of kid who told off his mother for throwing away a bottle instead of recycling it and had stopped eating tuna sandwiches *because the dolphins get caught in the net, and they die. I saw it on Blue Peter.*

'Look,' Mother said now, leaving the washing-up with some reluctance, untying her pinny so she could sit down. 'What's the worst that could happen? If Louis finds his father, Dan might refuse to see him. And that would be horrible for him, of course – but at least it would be real. It might help him get Dan down off the pedestal.'

Sky fidgeted with the tea-towel. The pain in her insides was very intense, so bad that it was difficult for her to form any words. 'If only,' she whispered, 'that were the worst that could happen.'

'What do you mean?' Skylark didn't reply. 'Oh, Lilian,' Mother said. 'You don't still think . . .'

But she still didn't say anything. The pain was too great. The pain of guilt, the pain of fear, the pain of love, and the pain of not being believed, not by anybody, least of all her own mother. The pain of doubting herself, then doubting the doubt, and on and on into infinity. When the pain opened up, there was no end to it. She knew it would go on for ever, until she was dead.

But being dead, she reminded herself firmly, was not an option.

Mother's clean, lemon-smelling hand was on her thigh. She was smiling, in that pitying, you-poor-deluded-thing way. 'He behaved awfully, dear. He wasn't the man we thought he was. He let us down.' She patted Skylark's thigh now, up and down with the nice clean lemony hand, pat-pat-pat. 'But that's all. He was a disappointing man,' she said firmly. 'Don't get swept up in your theories.'

She smiled briefly and stood up, retying the apron with the purposeful air of a knight putting on his armour to go back into battle. 'You must put that nonsense out of your mind, Lilian. That sort of thing doesn't happen in this country.'

3.

THE thing about it was that it had never felt right.

You know, don't you? When something makes sense. And when it doesn't, when there is something fundamentally puzzling and inexplicable and nonsensical about it, you also know. It's bothering, it's provoking, like a chair with three legs or a sentence with no verb or a jigsaw with a sky-piece missing.

I just don't understand, she had said in the early days, over and over again. These four words became her refrain, her endless lament, her bass drone. They played over in her mind as she washed up, as she poked mushy rice at baby Louis, as she tidied the kitchen, again. Whatever else she was doing – persuading Suze to give her her job back after maternity leave, pretending to be interested in Mother's codling-moth problem, lying awake in bed staring at the crack in the ceiling – was accompanied by *I just don't understand*. She never stopped searching. A thousand, a million times, she scanned the walls of the flat, emptied the cupboards and drawers, looking for anything, for the slightest clue.

He must have left something. Some trace. She had one photo of him, of him and Louis together, one that she had kept in her purse. She had given it to Louis to keep by his bed.

But there was no relief from it, not even when she was asleep. Just after he left, she dreamed she was standing on the top of a mountain, talking to God. *Why did he leave?* she asked Him.

He needed to learn the last letter of the Greek alphabet, God told her. *It was just something he had to do. It wasn't that he didn't love you.*

When she woke up, for a blissful three seconds she was convinced that she finally had a logical answer. Then she realised it was only dream-logic, and was crushed all over again.

And what about other people? Well, they were all very sympathetic. They rallied round with shopping and hugs and childcare. Aoife, who was pregnant with Franklin by then, bought her wine and chocolate once a week, because what else can you do? Said things like, *It's just different for men*, and *I always thought there was something off about him, I just didn't know how to tell you*. Mother saved the day by moving into the flat for a few months, to keep her fed and watered, until she found her feet caring for Louis alone.

But then, as time went on, they all grew understandably tired of *I just don't understand*. The reality was that nobody else in the world, not one single other person, shared her confusion or her grief. The truth of why Dan had left did not, fundamentally, matter to anyone else like it mattered to her.

As time passed, weeks, then months and years, the tune sung by the friends and relations began to change. *You just have to accept it*, they sang, in many different variations. *Get on with your life. Fuck him, Sky, you're a survivor. It's time to move on.*

She listened to their song, liked the song, understood why they were singing it. She even tried to join in, she really tried her best. But with the best will in the world, every time she opened her mouth, out came the drone: *I just don't understand*. So she shut her mouth, sang it silently, to herself.

The main thing she learned from Dan's disappearance was that we all, in some essential way, face life alone.

4.

OF course he had taken the yellow box with the birth certificate in it. That surely must have been the first thing he took. He would have made sure of that, before he bothered with the dumbbells and the bags of receipts, the screwdrivers and drill-bits, the photographs and underpants and the video of Aoife's wedding, in which he appeared in the background for approximately 0.5 seconds. He would obviously have taken the box before he ripped the page from her address book.

This almost total absence of any material reminder of his existence had a strange effect on her memories. Very quickly they lost their solidity, became flimsy and unreliable. This was strange, because four years is a long time to spend with one person. You'd think four years would sink right in, have some deep foundations. But without any tangible thing to hold on to, all that time came to feel as insubstantial as a bamboo house on long stilts. There was no ground beneath it, so the house swayed around in the breeze, this way and that, just one light puff away from crashing to the ground.

She had so many doubts and questions.

Had he loved her?

Why had he lied?

Who was Daniel Keller?

Was it possible that she had, in some way that she couldn't explain or understand, imagined Daniel Greene? Had she somehow projected him onto reality, to answer her need for stability and protection? For need, she thought now, was actually the word to express the connection between them. They had needed each other. It wasn't love. Love was something different: it was one and the same as truth, incompatible with hidden things.

She wanted to hold his birth certificate again, to settle this flimsy and ungrounded feeling. If she could just feel that real and indisputable piece of paper in her hand, the reality of it between her fingers, she would have at least something definite. So one afternoon when Louis was very small she opened the telephone directory and dialled the number for the Family Records Centre.

'Good afternoon,' she said to the grey-voiced man who answered. It was important to her to sound normal, educated and respectable, not like the kind of crazy oddball who would go around making up the father of her own child. 'I'd like to order a copy of a birth certificate, please.'

'The name, please?' he responded.

'Daniel Henry Greene,' she said. 'Date of birth, 14 October 1969. Registration district, Barnsley.'

'Bear with me,' said the grey voice, and then there was a click, a few minutes of the *William Tell* overture played on a synthesiser, followed by another click. 'Thank you, please allow ten days for delivery,' said the voice.

'Okay, hold on a second,' she said, heart hammering. 'So you definitely have it?'

'We have all birth certificates issued in this country,' said the man.

'Including this one?'

'I'm looking at it now,' he replied.

And, sure enough, it arrived the following week in the post, a photocopy of the document issued at the same time as the one she had once found in the yellow box. It was, in every detail, the very same.

Name and surname of father: Sidney Walter Greene.
Name and maiden surname of mother: Dorothy Greene formerly Ellis. Occupation of father: Coalface electrician.

And then one detail she had forgotten: it gave an address in Stainwell, Barnsley.

5.

THE morning after Louis's birthday, one coffee down and with the beginnings of a prosecco headache, she sat with him in the kitchen. In front of them was the box in which she kept everything she had that related to Dan. It was a small, cardboard container, just a bit bigger than a shoebox, and it was only half full. 'Is that all you've got?' he said, when he saw it.

'Yup,' she said. 'He took pretty much everything. He was very thorough. One thing we can say for sure about your dad, he didn't want to be found.'

'Let's just start off by not being sure about anything,' said Louis, quietly, and she marvelled again at the composure of this calm and forceful young person. He took out the top thing in the pile, the postcard she had received just a few days after Dan left, the last thing she'd ever got from him. He'd promised a letter, but when it arrived it was just this, a few words dashed down on a cheap postie.

Louis read it out loud.

Sky, love,

I'm okay. Just wanted to let you know. I'm in Spain, staying with family. I miss you and Lou so much and dream of the day we can be together again. D x

'Still dreaming!' She laughed with a bitterness that didn't suit her.

'He says he was really missing us, though. He wouldn't just say that, if he didn't mean it, would he?' Louis flipped it over. On the picture side was a flamenco dancer in a red frilled dress. The postmark: Andalucía, 6 August, 2001. He put it down and scratched absent-mindedly at his arms.

'I would have been a year and a half,' he said.

He had often searched his memory for images, feelings, anything that linked him to his father, but there was not much to go on. He had been too young. Sometimes he thought he could remember being held, a scratchy jumper against his cheek, but that was probably because Dan was wearing one in the only photograph he had of them together, the only one he had left behind. (*How come there aren't more?* he asked Mum. *We didn't take so many photos*, she replied. *You know what we used to remember things? Our brains. We still used our brains, back then.)*

Louis had always kept that photo by his bed. Dan was sitting on the green velvet chair in the living room, the one Mum only got rid of last year, when one of the springs poked through the seat. He had Louis in his arms, a few weeks old, wrapped in a white blanket, his baby eyes blinking open, tongue sticking out. Dan had his face tilted downwards, away from the camera, so his features were in shadow. Even here, pinned down in black-and-white print, his father somehow eluded him.

Louis looked at the postcard again. The writing was small and neat; controlled was the word that came to mind. *Staying with family . . .* Mum had told him Dad had family in Spain, that he had lived there, when he was young.

'Do you think he's still there? Andalucía?'

Mum shrugged. 'Who knows, Lou? He could be anywhere.'

'Did you ever go and look?'

'I never had the money. And you were only tiny. I was just about managing to get my socks on in the morning, back then, let alone getting myself on a plane.'

He put the postcard to one side and took the next thing out of the box. It was a printed flyer for the 'Street Party '96'. The letters were hand-drawn over a black-and-white photograph of a young woman he took a second to recognise. She was smiling, shimmying, caught mid-limbo. It was Mum, of course, wearing leggings with a hectic pattern on. He had never seen her face like that, not worried or tired, just radiating joy. Underneath the picture was more text clipped from newsprint. 'Saturday 13 July. Meet 12 noon Broadgate (Liverpool Street station).'

'That was our first really big party,' she told him. 'Here, watch this.'

Mum reached for her battered laptop, then typed 'M41 party' and the date into Google; top of the search results was a grainy, homemade-looking documentary. The footage was pixilated and the colour washed-out. It started with a man painting a sun face onto tarmac, a soundtrack of the kind of dance music Mum still listened to when she was washing up sometimes, two girls dancing in full body paint and flowered top hats. A crackly voice-over, a man with a posh accent. Louis recognised the voice.

We had invited people to meet at Liverpool Street station. This was because it was a main line station, with four tube lines connected to it. As far as the police were concerned, we could have been going off in any direction . . .

You know who that is? asked Mum.

'Uncle Rev.'

She fast-forwarded through a few more minutes:

people were running, whooping past a line of police; a group was dancing on top of a yellow lorry, stacked with speakers. And then she clicked pause, on a shot of a woman wearing a wig and a giant ballgown.

'There's Aoife, Franklin's mummy.'

She inched the footage forward, frame by frame. The woman who was Aoife in a former life moved regally past the camera, and as she did so, the bottom edge of her skirts caught the breeze.

She clicked pause again, backed up a couple of frames.

'And there he is.'

Hunched beneath the dome of skirt, was a shadow. It was a man, crouching, holding something in his hand. He had a pair of industrial goggles pushed up on his forehead, but despite that Louis made out the familiar square jaw, the long, dark hair.

You look like him.

Louis pressed his finger against the image, casting a flickering rainbow across his father's face, stroked it gently downwards.

But Mum had moved on. She was taking something out of the box.

'This is his birth certificate,' she said, handing it to him. 'I sent off for it, after he left. See the address on it? Stainwell? You've been there once. But I guess you probably don't remember the day we went to visit Granddad.'

6.

THEY had got there in the early afternoon, arriving on the bus from Barnsley town centre. Louis was starting to grizzle; it had been a long trip, four hours from King's Cross. Not a very nice day either, cloudy and muggy and close. Louis was two, and didn't always have his midday nap any more, especially not when there were trains to look at, so she had spent the whole journey keeping him happy with bits of apple and his favourite book (*Dig Dig Diggers!*) and wobbly walks up and down the train carriage.

The buggy rattled along the uneven pavement. Stainwell had once been a mining village, but was now a suburb of the larger neighbouring town. The bus stopped outside a grand village hall, and the high street was all solid grey stone buildings. But the streets on the outskirts were much like those in Henfield: the same wide, quiet streets, little 1950s two-up-two-downs with pebbledash walls, boxy places, like a child's picture of a house, four windows and a door and a path up the front garden.

Now they were near, so close to where Dan had grown up, and she gave Louis another apple slice to keep him quiet. He threw it on the floor, and ramped up the grizzle to a cry.

'Sorry, Lou,' she said. 'Not long now, promise. Soon we'll be at Granddad's!'

She hurried along. Sweat gathered in her armpits and soaked into her sad grey shapeless T-shirt. All the colour had drained away, out of her clothes, out of her life. All she wanted to do, all she ever wanted to do, was curl up and sleep. But she knew she had to try this one thing, the address on the birth certificate, even though Sidney had probably moved a long time ago. What were the chances of him still living in the same house after thirty-nine years? She focused her mind and rattled on, past a small, tarmacked playground, past a corner shop.

Some of the front gardens were full of junk – old prams and car parts and overflowing bins. Others had been brightened up with red or white geraniums – always geraniums – in pots. A few were more inventive: she pointed out to Louis one in which dozens of garden gnomes had been arranged into gnome-scapes around a fountain and some rocks.

What would Sidney's house be like? She thought about

what Dan had told her. *The booze has wrecked him. He's not somebody I can have over for a cosy little Sunday lunch.* She remembered how his hands had trembled as he'd said this, how he'd pulled at his rollie and stared over her shoulder. *Adrienne got me out.*

The sign on the next street said Newnham Terrace and her heart beat faster. It was only a cul-de-sac. In a few seconds' time they would be there, outside the house, Sidney and Diane's house, the house in which child-Dan had lived. Whatever happened next, it would have something to teach her about him. She rehearsed what she would say when Diane – for some reason she imagined it would be Diane – opened the door.

Good morning, Mrs Greene. I'm sorry to bother you. You don't know me, but I was your stepson's girlfriend. We were together four years. This is Louis. He's your grandson, Diane. Can I call you Diane?

No. She should leave all that until later, when Diane and Sidney were both sitting down, with a cup of tea.

The thing is that Dan has left us, without telling us why. He wasn't in a good way before he went, and I've been so worried. I've been worried that he might – well. I came all the way here to see – to see whether you know how I might get in touch with him.

Still too much. She couldn't hit them with all that.

I just wondered if I could come in for a few minutes. I

323

*know Dan, and I'm worried about him. There are a few
things I'd like to ask you.*

Yes, that was better. A light touch. Get inside. Softly,
softly.

She rattled on past numbers 10 and 12, then stood,
breathless and sweating, in front of number 14. The front
garden was much nicer than any of the others. It had a
small neat lawn and a border with shrubs and big daisies
and forget-me-nots. Whoever lived here mowed and
pruned, went to the garden centre, brought their dahlias
in over winter to save them from frost.

The booze has wrecked him.

It didn't look at all like the house she had imagined
for Sidney. But perhaps he had left, long ago. She pushed
open the front gate and manoeuvred the buggy up the
path. There was a doorbell, with a sticker beneath it.

Sidney and Dorothy Greene.

Dorothy?

That was Dan's mum's name. His dead mum's. And
she had died decades ago, when he was a baby. Why
would her name still be on the door? And what about
Diane? Whoever lived here clearly cared about her house.
Why would she leave Dorothy's name on the door?

She wiped the sweat off her forehead and fluffed up
her sweaty hair. That name was really odd, felt wrong,
the clang of a discordant note. *I just don't understand.*

She rang the bell, and waited. Peered through the frosted pane on the door. Nothing. Silence. No movement, no darkening of the glass.

'Mummy!' Louis whined, reaching for her with his chubby arms. 'Out!'

She rang the bell again.

Sidney and Diane were not in. But that was fine, they could wait. After coming all this way, they could hardly just turn around and go home. Perhaps they had popped out to the shops. They could sit here, in the front garden, until they got back.

'Out, Mummy!'

'All right, Lou. All right, just let me think what to do,' she snapped. Shifted her weight from foot to foot. Uneasiness crawled up her spine, sending its itchy, thorny shoots out down all her nerve endings. Sidney and Dorothy Greene. That was really weird. It was all wrong.

She pushed the buggy back down the front path, and out through the gate. Stood on the pavement and looked at the house, while the house looked back at her, its blank windows and blank door shut against a dead weight of dust and unspoken things. The sky was slate-grey, and a crow cawed from some close-by tree. She would walk up and down the road a few times, she decided, try to get Louis to go to sleep, while she thought about what to do next.

'Out,' squirmed Louis. 'Out, Mummy.'

'Yes, Lou, I hear you. Just hang on.'

But then she was running, with tears streaking down her cheeks, the buggy bumping crazily before her. Her body was propelling her away from that house, like a magnet repelled by some opposing force, as fast as she could go, along Newnham Terrace and back along the main road, past the shop, past the playground, even past the bus stop. Running, running, she didn't know why or where, only that she couldn't stay there, that she had to keep running, that whatever that house had to tell her, she didn't want to know.

7.

'So you never met Sidney and Diane?' asked Louis, disbelieving. 'You've known for all this time where he lived, and you've never got in touch?'

'I had my reasons,' she said.

'Did you?' said Louis. 'Because that was nine years ago, and they might not even be there any more. They might have moved. They might be dead. You got so close, and didn't do it, and now they might be gone.'

His mother shifted in the kitchen chair. The headache was thudding in her temples. Perhaps what she needed was another coffee to stave it off. She got up, fumbled around for all the bits and bobs, but Louis's accusing stare followed her from the cupboard to the stove to the fridge and back again.

'What?' she muttered, turning around. And then, 'Okay, Louis, you know something? I have many, many faults, but I know myself, and I trust my intuition. And that day we went to find Granddad, I knew things weren't right.'

'But, Mum,' Louis sighed, exasperated, 'you didn't know anything. You just had a feeling!'

'I trust my feelings,' she said, clanking the coffee-maker, slamming the cup, 'with good reason. Because the fact is, Louis, you still don't know the half of it.'

'Well, tell me,' he said. 'What don't I know?'

With that the clanking and the slamming ceased and his mother became very quiet. She seemed to go inside herself, letting her shoulders sag and closing her eyes. She took a long breath and sighed it out.

'There's more,' she said, 'that I haven't wanted to tell you. Stuff that I really don't know what to do with.'

'What do you mean, Mum?'

She put her hands to her eyes and her shoulders were shaking. Louis went to her where she was standing, by the stove. His head reached her collarbone now. He put both of his arms around her, and they stood like that quietly for a while. As they stood there, she thought, *Perhaps I'm not on my own any more.* She took her son's hand, led him back to the table and they both sat down.

'Okay, Lou,' she said. 'I'm going to tell you, and you can make up your own mind.'

'You know,' she told him, 'that before your father left we were being watched?'

'You've told me that,' said Louis, 'yes.'

'Well, after that day when you and I went to Barnsley, it all started up again. Only now he wasn't here.' She paused, breathed. 'After we went to find Sidney and Diane, they put me back under surveillance. I probably wouldn't have noticed, if it hadn't happened before,' she said.

'*Who* put you under surveillance?'

'I don't know.' She shrugged. 'He always said it was the police. They looked like police.'

'So, what, you saw them?'

'Yeah, sometimes. Like, I went to the supermarket one day, and there were these two men ahead of me at the checkout. I noticed them because, well, it's not often you see two big guys shopping together, is it? And they had a pineapple in their basket, just one pineapple. It was so weird.'

'Weird,' agreed Louis, 'but I mean, people do buy pineapples.'

'Wait. Just wait, Lou. But then I went and did a couple of other bits, met Aoife for a coffee, and when I got back here they were sitting – exactly the same two guys – outside the estate, on a bench. I saw them twice that day, in completely unrelated places.'

Louis tilted his head, weighing this up. 'Are you sure that was surveillance?' he said. 'They could have been anyone.'

'A couple of times I got back home,' she went on, 'and someone had been in the flat. There would be, like, a smell or something. Things would have been moved. Once they left my shoe lying on the floor when I knew I'd put it away. One day I got back and there was a bird in the living room. A crow. Flying about, flapping at the window. The door to the patio was banging open. I didn't leave it open, I just know I didn't.'

Louis was still looking unconvinced. 'Did you tell anyone, at the time?'

'I told Rev and Granny, and they both thought I was paranoid.' She glared hard at Louis, who was looking down and fiddling with his nails. 'Just like you do now, I bet.' He raised his hands, fending that one off. 'Okay, but get this.' She leaned forward. 'I knew that something was up. I just didn't know what. I got so strung out that I wasn't sleeping, at all. I had to go to the doctor and get on those tablets that I'm still on now.'

'I thought those were for headaches.'

'I told you they were for my head, Lou. Not for headaches. Anyway. One day I'd taken you into town. We went to see the changing of the guard.'

'I remember that! We went with Franklin.'

'That's right. And on the way back we had to change buses, and when we got off I looked up and saw we were right outside the Family Records Centre in Clerkenwell.

That's the place where they used to keep all the records, Lou. Births, deaths, marriages. Everybody who is born or dies in this country, there was a record of it there. It's where I got the birth certificate from, too. Anyway, I don't know what it was but as soon as I saw the sign I just knew I had to go in. I told them I wanted to look through the death certificates.'

'Were you looking for Dorothy? Because her name was still on the house?'

'No,' she said. 'I wasn't looking for Dorothy.'

'So whose death were you searching for?'

'Here is what I found.' She took the next piece of paper out of the box, unfolded it, and pressed it to her chest. 'Are you ready, Lou?'

He nodded. She handed it to him, and he again read it out loud.

'"I hereby record the death by natural causes,"' he read, '"of Daniel Greene, son of Sidney and Dorothy Greene, of 14 Newnham Terrace, Stainwell, Barnsley, South Yorkshire . . ."' Louis stopped. He looked at the paper again. He swallowed, and went on: '". . . on the thirtieth of September,"' he read, '"1970."' He swallowed again, looked up. 'I don't understand.' And then again, 'I don't understand, Mum. Whose death certificate is this?'

His mother reached out, took the paper from him and

put it on the table. She took both of Louis's hands between hers. She leaned over and kissed her son on his forehead, then leaned her own forehead against it. She whispered: 'Daniel Greene died, darling. Before he was even a year old.' Then she sat up and wiped her eyes. 'The boy who was born when your father said he was born, to Sidney and Dorothy Greene, in Barnsley, South Yorkshire . . . he never grew up.'

'So who was he?' said Louis. 'Who was Dad?'

She smiled, the saddest smile. 'That's it, Lou,' she said. 'That's it. I have no idea.'

James Calman: Good morning, Daniel.

Daniel Keller: Pleased to meet you, James.

JC: What would you like to tell me about why you're here today?

[PAUSE]

DK: They've finally agreed to cough up for this. Ten years too late, but there you go. Better late than never.

JC: 'They' being the police service?

DK: The Metropolitan Police.

JC: And you are still employed by the Metropolitan Police, is that correct?

DK: Yes. But I'm in a different unit now.

JC: Which unit?

DK: I moved into counter-terrorism. The Muslim Contact Unit. Several of us made that move, from the Special Demonstration Squad. You know, after 9/11. When everyone suddenly forgot about domestic extremists, and made Muslims the enemy instead.

JC: But it's the SDS that has agreed to pay for these sessions – why?

DK: [LAUGHS] Because I've spent almost ten years

throwing everything I had at them to get them to put me on long-term sick. Asked politely, at first. Explained my situation, how badly working for them had fucked everything up for me, my marriage, my kids, my head. Then, when they didn't give a monkey's about any of that, I threatened to sue them, and to tell the press everything I know about their operation. And – well. That did the trick, in the end. I'm on indefinite leave now, and the Met agreed to fund these sessions.

JC: To go through all that you must have felt very strongly that you needed psychological help.

DK: That has been obvious for a long time. I have not been a well teddy.

[PAUSE]

But now I'm finally here I don't know where to start, to be honest. After all that, I don't know where to begin.

JC: Perhaps you could tell me a bit about any symptoms of mental ill-health you experience, on a day-to-day basis.

DK: I don't sleep. There, that's a start.

JC: You don't sleep at all?

DK: An hour or two, here and there. Nothing you would call an actual night's sleep. Haven't had one of those for ten years.

JC: You've gone ten years without sleeping through the night?

DK: That is correct.

[PAUSE]

JC: And what is going on, in your mind, when you can't sleep?

DK: Fears. Thoughts. Memories.

JC: Memories of your time working undercover?

DK: Yes.

JC: What kind of memories?

DK: I run through scenarios in my head. People I knew. Things I could have done better.

JC: And fears?

DK: General fears, yes.

JC: And how do you feel, when you lie there awake?

DK: It's hell. My whole body is tense. It's like every muscle is locked. Adrenalin, I can feel it, just pumping around my veins. It's like I'm on high alert. Sometimes when I get up in the morning my jaw aches, because I've been grinding it all night long.

JC: I'm sorry.

[PAUSE]

DK: It makes me want to not exist, to be honest. It makes me want to die.

[SILENCE. SOBS.]

Sorry. I've never talked about this before.

JC: There is no need to apologise here. You can bring any feeling to these sessions.

DK: Okay. Thank you. I don't know. It's – well. It's all just such a mess in my head.

JC: Your feelings are bound to be complicated.

[PAUSE]

When you do fall asleep, even momentarily, do you dream?

DK: Sometimes. Yeah, I had one last night. I was somewhere beautiful, in a hotel, looking out over the sea. But the country I was in was a war zone. Like, I think it was Lebanon.

JC: What are your associations with Lebanon?

DK: It was where they used to hold hostages, wasn't it? Back in the 1990s.

JC: So were you being held hostage, in the dream?

DK: Not at first. I was in this beautiful hotel, looking out over the sea. It was night-time, and the sea was completely calm. But as I was watching, a plane came over and started dropping bombs.

JC: Dropping bombs on the sea?

DK: Yes, on this calm water. So then I had to leave the room and I was running through the hotel, looking for somewhere to hide. And I found a metal cupboard, like a locker. A metal locker. And I got inside, inside this tiny space.

JC: You shut yourself inside a metal locker.

DK: I had no choice. I knew they'd find me.

[SILENCE]

JC: And how do you think it would feel to open the door of the locker?

DK: They're looking for me. The soldiers.

JC: It would be dangerous?

DK: Of course.

[SILENCE]

JC: And when you say that at night you want to die, to not exist, do you actually plan to kill yourself? Or is it just a thought?

DK: I wouldn't actually do it. I mean, I have kids.

JC: So you have to stay alive, for them.

DK: Yes. If it were just for myself, I would do it.

[PAUSE]

           But even though Adrienne – she's my wife – and I aren't together now, haven't been for a long time, I'm still the breadwinner. [LAUGHS] Just about. I've hung in there.

[SILENCE]

JC: Some of these symptoms – the anxiety, the insomnia – are likely to indicate post-traumatic stress disorder, Daniel. It's sadly common after deployments such as yours. What we're going to do in these sessions is – we're going to try to open the locker.

DK: I don't know if I can do that.

JC: You might find that it's safer than you think.

[PAUSE]

JC: Whatever happened on your tour, you were following orders, Daniel. Doing your job.

DK: The problem was that the job was my life. Or, rather, it was my job to mess up my life. That's wrong, isn't it?

JC: Yes. It's wrong. You can say that, here.

DK: [SOBS] Thank you. Thank you.

8.

S HE swam up into consciousness from an anxious dream. For ten years now almost all of her dreams had been about searching: chasing along narrowing corridors and forcing herself through tiny constricting tunnels, following something or someone she knew was there, but couldn't quite see. In this one she had been in a field of some tall crop, corn or sugar cane. Pathways led her in circles; the stalks were sticky and tore gashes in her bare arms and legs. She searched in vain for any landmark or sense of direction, but the sky above was blank and white, like an eye with no pupil.

She rested for a moment, her pulse racing, in the red space behind her eyelids. The morning sunlight was of an unfamiliar brightness; she wasn't at home. Where was she? She opened her eyes to see a clean white ceiling, with a fan hanging in the centre. Of course, they were here. The guesthouse in Jerez de la Frontera. She and Louis had arrived last night, tired from the early start, the endless airport security, and the budget flight.

They'd found this place on lastminute.com. It was cheap but clean. The room had two narrow beds and two tall windows with slatted blinds. Louis was sprawled on the other bed, mouth open, his sheet on the floor. Knackered, poor love. She got up quietly and went over to the window, swivelled one of the slats to peer outside.

The guesthouse overlooked a square plaza bordered by trees. There were stone benches beneath each tree where she immediately knew, without ever having seen, that wizened old men wearing wide hats would play some kind of game, dominoes, perhaps, or bowling. Did they bowl in Spain? Or play backgammon, maybe? She had a memory of backgammon from the Encuentro.

The buildings around the square were low and white-washed, like everything in Jerez. They'd taken the bus there from the airport yesterday, along the bare scrubby coast, past olive and avocado groves, and up a twisting mountain road. The town perched at the top, a pointy white hat on the head of the hill. They had walked from the bus stop, bumping the suitcase over the cobbled streets, past cafés with open fronts, people sitting at tables smoking and sipping tiny glasses of what she told Louis must be sherry.

'It's like being in a film,' he had marvelled. This was his first time abroad. 'I don't feel like I'm really here.'

She knew exactly what he meant. She had forgotten, in many years of broke non-travel, the surreal sensation of stepping onto a plane and then off a plane to find yourself *somewhere else*.

On the bed behind her now, he was stirring. His face was creased and his dark hair sticking up in a way that melted her heart. 'Morning, love.'

He yawned. 'Oh, wow,' he said. 'We're still here. I thought I dreamed it.'

'Nope,' she said. 'We've really done it. I blame you.'

'Well, I blame you. For everything.'

This was only half a joke. He sat up, rubbed his eyes in the way that set off a million memories and associations. Dan felt so close at the moment. Not physically close – although that was, terrifyingly, a possibility – but present in her mind. Everything reminded her of him; everything posed the question.

'So are we going to do it this morning?' Louis asked.

Adrenalin fizzed and her stomach contracted. 'Oh, God,' she said. 'What do you think?'

Louis's face assumed that determined look. There was something hard about that look, she thought, something brittle and breakable. Her heart ached for him, its usual guilty worrying ache. 'I think we do it,' he said. 'As soon as possible. I don't want to think about it any more. Let's get it out the way.'

'Okay,' she said, and stopped herself asking what *out the way* even meant, in this context. Whatever happened when they went to Adrienne's house, getting this *out the way* would not be an option. Either they'd find nothing, and nobody, in which case they would be no closer to answering the questions they wrestled with every day. Or she would be there – maybe even he would be there – and a door would open and they would all find themselves together in a new and complicated space.

9.

THEY had got Adrienne's address from the marriage certificate. And they had got the marriage certificate after searching for her name online.

They had looked for him first, and come up with nothing; nothing had ever come up for him, in ten years of online searches. Just to be on the safe side, they checked for a Daniel Greene, the person they knew he was not. Obviously there were a million Daniel Greenes on Facebook and Twitter and LinkedIn, everywhere from Adelaide to Arkansas, but none looked like Dan. On the off-chance, they also tried Daniel Keller, just in case the driving licence had been real in a way that everything else wasn't. Again, thousands of results, thousands of photos, none of which were him.

'The problem is, it's just such a common name,' moaned Louis, in frustration. 'If only he was called Humperdinck, or something.'

'Hang on,' she said suddenly. 'I have another idea. He

used to talk about an aunt. She was really important and special to him.'

'Oh, yeah? Was she called Humperdinck?'

'Adrienne,' she said. 'I don't know her surname. Obviously he might have made her up, too. But there was something in the way he talked about her . . . Let's just try.'

His fingers hovered over the keyboard. 'So Adrienne Greene?'

'No,' she said. 'That's not his name, it's just not. Try Adrienne Keller.'

Louis typed it in, and a gallery of pictures appeared on the screen. Smiling women, serious professional women, some young, some old, some profile shots, some on holiday, some drunk and hugging friends. They scrolled through, looking for a sign, any sign at all, that this was the Adrienne, aunt of Dan. She'd be an older lady, presumably, of the grandparent generation. Tanned from a life in Spain. Perhaps too old to be on social networks.

'Hang on,' she said suddenly. 'Scroll back up.' She pointed at a Facebook profile picture: it wasn't a photograph, but one of the illustrations from *Alice's Adventures in Wonderland*, the one of Alice pulling back a curtain. She had a funny feeling about it. Had Dan said something about that book once? 'Look at this one,' she said.

Louis clicked on it. The profile wasn't public. All they could see was the picture and the name, but underneath the name it said Jerez, Andalucía.

'He told me she lived in Jerez,' she said, triumphantly. 'This is her. It has to be. There can't be two Adrienne Kellers in Jerez. It's not even a Spanish name.'

'And look,' said Louis, pointing at the screen. 'Barnsley High School, class of 1987. It must be her. That makes her the same school year as Dad.'

'That can't be right. She must be older.' Sky frowned at the screen. 'It doesn't make sense.'

'Shall we send her a message?' Louis hovered the cursor over the button.

'Stop, Lou,' she said forcefully, snatching the laptop and snapping it shut. 'Wait. Don't do anything yet.'

'Jeez, Mum.' Louis was irritated now. 'Why not?'

She put the computer down. 'Sorry,' she said. 'Sorry.' She rested her elbows on the table, and her head in her hands. She squeezed her eyes tightly shut, trying to contain whatever was bubbling up inside.

Louis shoved her gently with his shoulder. 'Are you okay?'

'Yeah,' she breathed. 'It's just— God, Lou, I don't know. I don't know if I can do this.'

They sat there side by side for a moment. Outside the kitchen window a plane tracked through the sky; sounds of the rattling football cage drifted up from the forecourt below. *I just don't understand.* The old refrain was playing as usual in her mind. It had been with her for so long that she couldn't imagine life without it. Whatever the truth was, did she want to know? Did she want Louis to know?

'Think about it this way,' said Louis. 'What's the worst that could happen?'

'You know, that's exactly what Granny asked.'

'But seriously,' said Louis, 'what are you worried about?'

She pulled herself together, gave him a weak smile. 'Probably nothing,' she said. 'Sorry, Lou. Just give me a bit of time.'

But the next morning, after dropping him at school, she went online again, on a hunch. The Family Records Centre had closed a few years back; now all the birth, marriage and death certificates were online, like everything else.

She entered her payment details and chose the 'marriage certificates' option. Typed in *Daniel Keller* and *Adrienne*. Both born in 1969. Married in the early nineties. The loading bar ticked over for a few moments, and then there it was: Adrienne Keller wasn't Daniel Keller's aunt.

She was his wife.

Skylark leaned forward, pressed her forehead against the screen. He was married. He had been married, all along. When he went away, for work, had he been going home to her, to Adrienne? And he loved her, he had told her that, in so many words. She had felt it, sensed it, from the tone in his voice. That was why she had remembered so clearly the conversation they'd had about her. *Adrienne got me out. She's the only one who looked out for me.*

Had he been with Adrienne, had loved Adrienne, the whole time? When they met, when they first kissed, when he moved in, while she was pregnant, while she was in labour with Louis? What was she like? She lived in Spain – was she an adventurer? A Bohemian, even? Was she beautiful? She was beautiful, of course. Dan would never have turned the lights off with her.

Her stomach churned. Acid rose up her throat. Her palms and face were damp. The room around her receded and she was back there, alone, on the ice-cold Himalayan mountaintop. Part of her, of course, had been there all along. She had known in her bones, known something.

He had kept her in the dark.

He couldn't bear to look at her.

And then her body lurched and she only just made it to the sink, retching. Retching up breakfast, then bile, then just heaving, purging herself of what felt like a lifetime of poison. When it stopped, she carefully, carefully, lay down on the kitchen floor. The lino was cool and sticky. He had chosen that lino, and laid it. He had painted that ceiling, and he'd missed a bit, right in the corner. He was still here, fuck him. He'd left bits of himself everywhere. Left bits of himself inside her head.

You know, I love you, Sky. You and Lou.

Who would she have been, without Daniel Keller, without Daniel Greene? What would she have done? Would she have been braver, bolder, occupied her inner landscape more completely? Would she have explored those ancient forests, stood firm among the megalithic stones? Compromised less, risked more? What could the world-changing group have been? Would it have stayed truer to its original impulse, its flaming arrows of hope?

I just don't understand.

The invisible threads of his lies stretched from her heart right out into the far corners of her life. He had permeated and penetrated every bit of it. She had submitted to him in every way: her emotions, her politics, her body. Everything would have been different. She would have been someone else.

She wouldn't have had Louis.

And that was where it all broke down because her son was the one thing in life she had done completely right. She wouldn't change him, wouldn't change anything about him. Her mind shattered into fragments that spun off into spirals, into fractals. It was a puzzle she couldn't piece together, a sentence with no verb. *Tick-tock, tick-tock*. Her life was passing, and she couldn't get out, would never get out. She was trapped in a dream that wasn't even her own.

10.

THAT afternoon, she visited Rev in his new studio. It was just off York Way, near where it had been back in the old days, except the whole area had changed beyond recognition. The bus stop-started along City Road, past the pedestrianised plaza where the arches had once been, past the station entrance where there was no longer a newsstand. The drinkers, junkies and prostitutes had vanished; the chewing gum had been power-sprayed off the pavements. King's Cross was now the 'gateway to Europe', a place where things were simple and clear, where well-dressed people walked briskly in straight, efficient lines.

York Way was a spiky, jangling panorama of drills and cranes, men in fluorescent tabards and hard hats busily pulling down the grimy old bricks and replacing them with reinforced concrete, gleaming steel and glass. The petrol station by the canal was now a waterside cocktail bar. The gasworks still loomed, remnant of a messier past, but everything around it had been scrubbed and streamlined and simplified.

It was, she thought, as though the more complicated and confusing the world became the simpler and clearer it needed to appear. Reality was whatever people wanted and needed to see. Looking out at that reconstituted view, she felt herself once again on that distant remote and unreachable mountaintop, completely alone.

She got off at the canal and turned down a side-street, where she stopped in front of a smooth metal gate in a concrete wall. Rev had bought this place a couple of years ago. He was loaded, now, his machine beasts sold to collectors all over the world; they were auctioned and re-auctioned. But if she ever teased him for being a sell-out, he said, *Well, it helps pay your rent, darling*, which shut her up.

She pressed the button on the video intercom. One of Rev's assistants – he had many assistants now – answered and the gate hummed open. She stepped through into a concrete amphitheatre, through which a set of steps led up to the glass studio doors. Rev was leaning over a table with a young, pink-haired woman.

'Sky,' he said, when she walked in. 'Jesus, you look awful. Here, sit down. Hannah, could you get us a jug of water?'

'Of course,' said the pink-haired girl, and hurried over

to the slate-and-marble kitchen area. She set a jug and two glasses on the table, then Rev gave her a little nod and she scurried away.

'Has something happened?'

Skylark poured herself a glass of water and drank it, its coldness bringing her back into her body. 'He was married,' she said.

'Who? Oh,' said Rev, instant weariness creeping into his expression. 'Him.' She nodded, and he put his hand over hers. 'Skylark McCoy, you're not going down that road again, are you? I thought you'd decided—'

'It's for Louis,' she said. 'Louis wants to find him.'

'Ah.' Rev leaned back and ran one hand over his shiny bald head. 'I see.'

'We started digging around on the internet and I looked up a woman Dan told me about, ages ago. He always said she was his aunt. But we found her online, and they're the same age. I've ordered the marriage certificate. Adrienne Keller and Daniel Keller. The same name I once found on his driving licence.'

Rev leaned back, exhaling. 'You hear about these men who live double lives, but you never actually think—'

'But it wasn't just that,' she said. 'Don't you see?'

Rev put both hands carefully in front of him on the table. 'See what, Skylark?' he said.

When she had told Rev and Mother, years ago, neither

of them had believed her. Both of them had thought she was mad, that it wasn't possible. Rev thought Dan was just power-hungry, a sociopath, a liar and a cheat. Mother had insisted, *That kind of thing doesn't happen in this country.* Talking to them about it made her feel like she was going mad, so she had given up and packed her suspicions away in a mental box marked 'Private'.

'He was a spy,' she said, now. 'Or an agent provocateur. He was sent to disrupt the activities of the world-changing group, to derail the movement and protect the capitalist system. And I was his cover story. We – me and Lou – were his cover story.'

Rev listened, his face serious. This time, he did not tell her she was mad. 'I think it's really important to try to separate what you know,' he said, 'from what you fear. What do you actually know?'

'I know he lied about his name. And that he had a fake birth certificate in the name of a dead child.'

'Right.'

'I know he was married to somebody he told me was his aunt.'

'Okay.'

'I know that shortly before, and also after, he disappeared, I was being watched.'

'Do you know that?'

'Yes.'

'Who were you being watched by?'

'He told me they were police.'

'But if he was a spy, why would they have needed to watch you? And who was he working for? The government?'

'Perhaps. Or big business. The oil industry.'

'Or something in between, like—'

'Some secret part of the state. Where the state meets big business. The blurred boundaries, the inner workings of the capitalist machine.'

Rev nodded slowly. 'I'm not saying you're wrong,' he said eventually. 'I just think,' he turned his palms upwards, 'you may never know. It may be best to accept – to try, as far as possible, to get on with your life.'

'But don't you see?' she said. 'I can't. I will never know what my life is, what it means. And Lou—' She stopped. Tears ran down her face, and she wiped them away with her sleeve. Rev took her other hand gently and held it. 'Louis,' she whispered. 'What shall I tell Louis?'

'I can't tell you that,' said Rev. 'All I can say is, you're his mother. Trust your own instinct. How important is the truth?'

11.

Louis put down his after-school toast and peanut butter, and stared at her. 'I didn't understand any of that,' he said.

'Right,' she said. 'Do you know what a spy is?'

'Kind of,' he said, and then, 'actually, no.'

'Okay,' she said. 'So you know Lord Voldemort? When he inhabits Quirrell's body? That's kind of what I mean. Somebody who pretends to be someone else.'

Louis's eyes filled with tears. 'You're saying Dad is like Lord Voldemort?'

'I don't know,' she said. 'But this morning you asked for my worst case scenario. And that's it.'

'But,' said Louis, wiping his eyes, 'you told me he was fighting for a better world.'

'Maybe, darling. Maybe he was. Or thought he was.'

'But if he wasn't . . .' said Louis '. . . that means I'm half Voldemort.'

'No,' she said passionately, gathering him in her arms.

'You must never think that. Whatever Dad was, why ever he came and why ever he left, he had his reasons. He wasn't a bad person. He might have just – become confused.'

Louis struggled out of her hug. 'But if he was Voldemort, or even just a Death Eater, then he was lying to you,' he said. 'And lying is wrong.'

'Yes,' she said. 'Yes. But—'

'And Adrienne wasn't his aunt, she was his wife? So he has another family, maybe? And he was lying to them, too?'

'He probably didn't—'

'No, Mum,' said Louis. He stood up. Tears had gathered in his eyes. 'Don't make excuses. If he was lying to you, and to Adrienne, if he was lying to Uncle Rev, if he was lying about wanting a better world . . . if he had me because of a lie, and then left because of a lie, he's just Voldemort. That's all.'

She nodded, bit her lip.

'And if you knew he was lying,' Louis went on, 'if you've known that all along, then you've been lying to me, too.'

'No, Lou, no. I would never—'

'Yes, you have!' Louis shoved one of the kitchen chairs, hard, and it tipped and landed on the floor with a crash. 'How do I know if I can believe you now? And what about me? Who am I? I don't even know!'

He ran out of the kitchen and she heard his bedroom

door slam. She had dreaded this moment for years, but now it was here, she felt nothing. Numb. She had tried to do the right thing, tried to do her best for Louis, but it was too late. It had been too late, ever since he was born. Because, like everything else, their relationship was built on foundations that weren't solid or real. It was part of the bamboo house, swaying in the wind, forever threatening to crash to the ground.

The football cage rattled. Somebody walked past the kitchen window, on their way to the lift, casting a fleeting shadow. From Louis's bedroom came the muffled sound of sobbing. She sat there, numb, with the computer in front of her.

After a few moments she stood up and went into Louis's room. She sat down beside him on the bed, and put her hand gently on his back.

'You're right, Lou, and I'm so sorry,' she said. Then she paused. 'But I promise you, from now on I will always tell you the truth.'

He sniffed. 'Even the bad bits?'

'Yes,' she said. 'Especially those.' She leaned over to hug him and, to her great relief, he let her. 'And if you still want to find him, I'll do everything I can to help you, okay?'

'So will we go to look for Adrienne?'

'If you want to,' she said. 'Of course.'

12.

STRAIGHT after breakfast, she showed the address they had got for Adrienne Keller from the Spanish electoral roll to the grey-haired woman who ran the guesthouse. She looked at it, puzzled, then typed it into her computer. She turned the screen to show a blue dot in the middle of a square of green.

'Is not in the town,' she said. 'Is, how you say?, nowhere.'

'Can we take a bus?' she asked.

'No, bus no.'

'A taxi?'

The woman shrugged. '*Quizás.*'

The town was just waking up as they left, shutters clattering up on the cafés and shops. There was a large minibus-taxi waiting in the square, and she showed the address to the driver, who was wearing shades and a dusty Yankees cap. He looked at it and nodded. '*Sí.*'

'You know it?' She mimed, trying to remember her guidebook Spanish. '*Sabes*?'

'*Sí, claro.*'

'How much? *Cuanto*?'

He shrugged, and pointed to the meter. 'It doesn't matter, though, does it?' said Louis, impatiently, and opened the door.

The cabby drove them back down the hill but, instead of taking the coast road, turned inland, then off the tarmacked road onto a dirt track, tutting and snorting as the minibus bumped and jolted over dust and rocks. It was still early, but the sun was already getting hot, and her legs stuck to the tacky plastic seats. She felt oddly calm. Louis had his eyes fixed out of the window; neither of them tried to talk. She took his long soft hand, and he didn't pull it away.

'*Aquí,*' said the taxi driver, stopping at the top of a steep smaller track. '*No puedo más.*'

'I don't think he can drive down this bit,' said Louis.

She mimed to him to wait, and they got out. Walked down the hill, hand in hand. The earth was yellow, the stones were yellow, and the foliage that climbed and scrambled on either side a bright dry green. On their left in a clearing there was a small, deep pond, which must have been used for swimming, because somebody had built a little diving platform. Crickets chirruped everywhere, but there was no human sound. At the bottom of the hill the track opened out and they saw the house.

'Look at this place,' whispered Louis.

It was a low wooden building. The red-tiled roof extended over a veranda, with a rough wooden table underneath. Two wooden benches looked out over a garden filled with swaying avocado trees. The beams supporting the roof were made from rough, twisted tree trunks, so the whole place was wonky and higgledy-piggledy, like a house from a fairy-tale.

This is the kind of place I want when I'm old. A little house where I can sit and watch the sun set.

'There's nobody here,' said Louis, who was examining the shuttered windows, the bolted door. 'It's all locked up.'

'*Ya se fué.*' A child was standing at the top of the track, dressed in swimming trunks and flip-flops.

'*Buscamos Adrienne,*' said Louis, who had practised this phrase in the plane on the way here. '*Donde está?*'

'*Adrienne se fué,*' said the child, with a puzzled expression. '*Ayer. Con los gemelos.*'

'They are gone. Left yesterday.' The taxi driver had materialised beside the child. 'She left with her— How you say? With her twins.'

They had twins. Louis had siblings.

'*Se fueron con tremenda prisa*,' said the child.

'They left in a big hurry,' said the driver.

'*A Inglaterra*,' said the child, then broke into a smile. 'Adrienne to Eeeeeengland!'

JC: You seem agitated.

DK: Yes. Yes, I am.

JC: Has something happened?

DK: Adrienne and the twins have been moved back here, from Spain.

JC: Oh?

DK: The SDS advised that they were no longer secure. They have a system that flags up when anybody searches for the records of one of its former undercover officers or their relatives. So they've been moved to a safe house in England. Nobody can tell us whether they can go back.

JC: Because somebody searched for a record relating to Adrienne?

DK: That's right.

JC: Who?

DK: Somebody I used to know.

JC: One of the – what do you call them? – wearies?

DK: That's right.

JC: And how did this person know her name?

DK: That was – that was a mistake. My mistake.
        I used Adrienne's name – not her full name –
        as part of my legend. And she – this person
        – put two and two together.

JC: Only now? Ten years after your deployment?

[PAUSE]

        She must really want to find you, to be looking
        after all this time.

[PAUSE]

DK: I guess she does, yes.

[PAUSE]

JC: Why do you think you used Adrienne's name
        as part of your cover story?

[PAUSE]

DK: I don't know. It was a mistake. Obviously.

JC: Was there some part of you that wanted to
        tell the truth?

[LONG PAUSE]

DK: Maybe. Yeah, maybe.

JC: Was there a part of you that wanted this person
        to be able to find you?

[LONG PAUSE]

DK: No. No. I wouldn't say that. I did not – I would
          never – want to compromise Adrienne, or the
          kids.

JC: Not consciously, perhaps.

[PAUSE]

          And who was this person? The person you
          told?

DK: [COUGHS] Skylark. Lilian May McCoy. Her
          name is Sky.

JC: Did you know her well?

DK: I knew her well. Yeah.

[PAUSE]

          She was actually – she was . . . well, yeah.

JC: You had a relationship with her.

[PAUSE]

DK: Yes.

JC: A romantic attachment?

DK: Yes. It's something – one of the things – I
          feel . . . It was wrong. We weren't supposed
          to, strictly speaking, form attachments like
          that, obviously. Or, rather, we were tacitly
          encouraged to form them, as it helped us to

integrate with the group – we just weren't supposed to stay together. But . . .

[SILENCE]

JC: Did you fall in love with her? With Sky?

[PAUSE]

DK: At times I thought so, yes.

[PAUSE]

It's hard to – well, I find it hard to put a label on feelings. Like, love. I know I love my kids, no question. But as for women, even my wife . . . I don't know.

JC: Or was there some kind of dependency, then?

DK: Yes. Dependency, yes. I depended on her a lot, when I was in the field.

JC: And was that mutual?

DK: The dependency? Yes. She depended on me, too.

JC: Did she love you?

[PAUSE]

DK: I don't know. That wasn't really something we – talked about. Or maybe we did, once.

[SILENCE]

JC: That must have been difficult. When these feelings got out of hand.

DK: [LAUGHS] Well. It wasn't really difficult at the time. It was good, or at least it was good at first. She introduced me to a whole different way of life. To a kind of new version of myself.

JC: But knowing that you couldn't tell her the truth . . .

DK: Yeah. Yeah, that . . . It really messed with my head.

[PAUSE]

JC: And why do you think she's still looking for you, ten years later?

[SILENCE]

          Can you imagine why she might want to see you now?

[PAUSE]

DK: There was . . . Like I said, things got out of hand.

JC: Can you explain what you mean by that?

[PAUSE]

DK: At a certain point, she became . . . she got pregnant.

JC: Ah.

[SILENCE]

And what became of that pregnancy?

[LONG PAUSE]

DK: Sorry.

[PAUSE]

This is something I find it very hard to think about.

JC: Did she have the baby?

DK: [WHISPERS] She did, yes. Louis. [CLEARS THROAT] He's ten now.

[PAUSE]

JC: And do you – have you—

DK: I don't see him. Never have done, since the end of my tour. But . . . well. My old supervisor, Martin. He's not a bad guy, became a friend over the years. He still works in the Squad, has access to all the surveillance. He sends me a picture, now and again.

[PAUSE. SOBS.]

So I know he's – so I know he's okay. He looks like a sweet kid.

JC: So your supervisor knew about this?

DK: [SOBBING] He promised to keep an eye. For me. [SOBS] Sorry.

JC: There is no need to apologise to me. This is what we're here for. Please, have a tissue.

DK: Thanks.

[SILENCE]

JC: How have you been getting on with the medication?

DK: [BLOWS NOSE] Yeah. Okay. I mean, it zonks me out a bit.

JC: There can be side effects, at the beginning.

DK: But I have been sleeping. So that's great.

JC: Good. That's important. What we hope to do with the medication is to – open up some space and some energy that you can use to face and deal with some of the difficult issues in your life. Maybe things you've been avoiding for many years.

[PAUSE]

It's going to be important not to condemn

yourself to a life sentence, Daniel. Inflicting pain and suffering on yourself doesn't actually help anyone. Not you, not Adrienne or your children, not – what was her name?

DK: Sky?

JC: Not Sky.

DK: [SOBS] I just – I wish things were different.

JC: But they're not. That's important. Start where you are.

DK: I don't deserve—

JC: Daniel. All you can do at this point is try to get better. Only then will you be able to help anyone else involved in this situation. You're here, and that's the very important first step.

13.

*D*ANIEL *Keller.*

For the second time in her life, the name stopped her in her tracks. The hubbub of the party woozed and melted as she stared at the screen of her phone. It had pinged with a new email, and she'd taken it out of her pocket to check it wasn't Louis, some emergency. And it was there, the name, waiting for her, a message from across time, from across the lines.

Sender: Daniel Keller

Subject: Long time

Her heart pounded. What? Was this someone's idea of a joke? Not here. Not now. She stuffed the phone back into her pocket, knocked back the rest of her champagne.

'I heard that Hollywood guy is going to be here,' Suze was saying in an ostentatious whisper, scanning the room over her shoulder. 'What's his name?'

'Dennis Hopper,' she replied. She was glad to have Suze and her giant hot-pink jumpsuit to hide behind, as even before the email arrived she had been feeling discombobulated by the opening of Rev's show. He had been signed up by a new gallery in the West End, one with a flashing red neon sign outside advertising STREET ART. Inside, all the pipework and plaster were exposed, perhaps to make the gallery look like a squat, only this was a sanitised and manicured squat-simulacrum, carefully filleted of community spirit and political energy. People with expensive asymmetric clothing and haircuts milled about among the machine beasts, sucking from flutes of champagne.

It was, she had told herself firmly, fun to go out in the evening, without Louis, just herself, in a fashionable venue. She had put on her old silver jeans and a tight black top, and looked like some previous version of herself. But she also, on some deep and intractable level, felt completely out of place.

'What film was he in again?' asked Suze.

'*Easy Rider*?' she answered. '*Blue Velvet*?' but Suze looked blank. She had grown up in the Philippines and did not share many Western cultural reference points. 'He's a big art collector,' she explained.

'I know the name,' said Suze. 'I'm sure I'll recognise him when I see him.'

She felt a pre-emptive wash of weariness just thinking about how desperately Suze would try to ingratiate herself with Dennis Hopper, if he were to arrive. 'I don't think they're sure he's coming,' she said.

'Suze Rodriquez?' said a voice from behind them. Tom Cooper, *Evening Standard*. 'Are you enjoying the show?'

Suze turned around and became engrossed in talking to the reporter; Suze loved talking to reporters. Meanwhile, Sky looked around for somewhere quiet, where she could sit down and read the message. Whatever it was, whoever it was, she didn't want to find out here, in front of all these people.

'A drink, madam?' A waiter dressed in an immaculate tuxedo was holding out a tray.

She took a glass of fizz, her second. She took a sip, acid furring her tongue. 'Is there anywhere quiet around here? Somewhere I can sit down for a minute?' She gestured to her phone. 'I have a— There's a situation I need to deal with.'

'You could go upstairs,' the waiter said, pointing. 'Through there. It's marked private, but nobody will care.'

'Thank you. Thank you.' She took a longer glug from her glass, which seemed to have emptied very quickly, but the waiter had moved on, and she didn't want to chase him to get another. She made her way through the crowd, past the Private sign, up the stairs. Above the gallery she found a small, poky room with two black leather sofas, a coffee-table and a kitchenette. The babble of the smokers drifted up from the pavement below. She sat on one of the sofas and opened the email.

Dear Sky, I believe you have been looking for me, so I thought I would spare you any more trouble and get in touch myself.

Oh. My. God, she thought, putting a hand to her throat. Was it actually him?

As I think you are now well aware, I was not honest with you during the time we spent together. At the time I believed there were good reasons for that. It's hard to find the words for what I need to say to you. Words don't seem adequate. If you would ever agree to meet me, we could talk about it and perhaps start to move on from that difficult past. But I will, of course, understand if you don't feel able to do that.

It was him. Actually him. Was it, really? And who was he, anyway? Why was he writing now, after ten years? The questions grew and proliferated, uselessly, threatened to drown her, even as she sat on that plasticky sofa, at that plasticky party, life going on, as life always did, mercilessly, relentlessly. She needed another drink, more than she had ever needed a drink before. She closed the email, stood up, and went to find one.

Downstairs, the energy of the party had changed. People were on their second glass of champagne, or third. Cheeks were flushed, and voices were louder, excitable.

'What film was he in, again?' a woman in an asymmetrical dress was saying.

'You know, I can't actually remember,' replied a man who was too old to be wearing a baseball cap backwards. '*Apocalypse Now*?'

She wandered through the crowded rooms, trying to lose herself in the art. The machine beasts in this show were smaller, more contained: a hummingbird delicately carved from a tin can; a bush baby moulded from discarded shotgun cartridges. Rev had scaled down his work recently, made it more intricate and beautiful, less raw, less ragged, and she teased him that this was to appeal to rich collectors who wanted pieces small and pretty enough to put in their homes. *Still believe art only happens outside the market, Revvy?* She could see him in the far corner of the main space, a battered top hat on his head and a white shirt with a wide ruff, but she didn't go over: she avoided talking to Rev when he was in Famous Artist mode.

'Skylark?'

For a moment she didn't recognise the man standing in front of her, or the woman he had his arm around. He had short dark hair, a pot belly beneath his Ramones T-shirt, and the air of a faded rock star.

'Mikey!'

He smiled. 'It's been a long time.'

The woman with him had her hair in a wrap, and wore a colourful zigzag dress. Her face was calm but sharp.

'This is Emma,' said Mikey. 'Emma, this is Sky. I've told you about her.'

'Oh, yes,' said Emma, smiling and holding out her hand. 'Of course. Nice to meet you.'

She took Emma's hand and shook it with a bewildered kind of formality. Just by the way they were standing, close together, touching each other, she could see that Mikey and Emma were happy, that they were in love. Her throat was tight, her shoulders ached, as if she were carrying a heavy weight on her back.

'You look well,' she managed to say, and Mikey held up his glass, which was filled with orange juice.

'This would be why,' he said. 'Two years dry.' Emma smiled up at him proudly, protectively. 'Took me a while, but I got there in the end.' He looked her up and down with his old blatant appraisal. 'I heard you had a kid?'

'Yeah, Louis,' she said. 'He's ten now.'

'And your copper?'

She stopped. Her stomach lurched. 'Sorry?'

'The cop,' Mikey said. 'What did he call himself? Dan.'

Mikey's face with its glimmer of mischief and Emma's face with its patient smile hung before her, like mocking empty masks, while the rest of the room – the machine beasts, the plaster walls, the gleaming tall glasses, the miniature waiters – pulsed and receded. When she managed to speak, after what might have been seconds or minutes, her voice seemed to come from somewhere very far away.

'He left us,' she said, 'just after Louis was born.'

Mikey nodded, and took a sip of his juice.

'You knew?' She stared at him. Emma was still smiling, but the smile no longer looked easy and natural: it was a frozen smile, the smile of someone who is trying very hard not to stop smiling. 'Please tell me,' she said, in a harsh, ragged voice, 'what you knew.'

Mikey's eyes glimmered. 'Don't say you didn't . . .?'

The room pulsed again, or maybe it was her head that was pulsing. Either way, she didn't know if she could keep standing up.

'Sorry. I thought you would have heard by now. Seeing as it's all started to come out. Drip, drip, drip.'

Emma reached out and touched the top of Skylark's arm. 'Stop, Mikey,' she said. 'She's upset.'

Suddenly furious, Skylark shrugged off Emma's hand. 'No, it's okay,' she said loudly. She was distantly aware that other people were turning around, that they were pausing in their animated drunken conversations, that she was on the verge of Making a Scene. She leaned forward. 'Tell me, Mikey,' she hissed. 'Tell me everything.'

'Okay,' said Emma, softly, smilingly, 'let's find somewhere to sit down.'

Skylark followed Mikey and Emma back through the crowd, back past the Private sign, up the stairs. They sat next to each other on one sofa, arms draped around each other; Skylark sat on the other.

'So you haven't been in touch with anyone,' said Mikey, crossing his thin legs, resting a hand on his little pot belly, 'from the old scene?'

'I see people,' she said. 'I see Aoife and Rev. Everyone else kind of drifted away. Or maybe I drifted away,' she said. 'I'm not sure.'

She wished Louis was there, so she had someone to sit with on her sofa. She felt she would handle this whole situation much better if he was beside her.

'Yeah, that's it,' said Mikey. 'Because if you had, you would have heard.'

He leaned forward, put a Rizla in the crease of his jeans and filled it with tobacco. 'Turns out that back in the day there were men, lots of them,' he said, 'who joined various world-changing groups, like yours and mine, who hung around for years, who had girlfriends and sometimes even kids, yeah. They were pigs. Secret police.'

The room felt very hot, so she got up and opened a window. As the petrol-tinged breeze blew in from the street below, she realised she was calm, so calm now. This conversation felt inevitable, natural. Of course Mikey would be the one with the missing piece. Mikey, who had always been her gatekeeper, for better and for worse.

He lit his cigarette, which was definitely not allowed inside, not now. He threw back his head and blew smoke up at the ceiling, then met her gaze with his mischievous glimmering eyes. 'I mean, I assumed he was one of them. As soon as I heard. Call it instinct, call it a long time on the scene. He just smelt like a pig to me. Always.'

He pulled a phone out of his pocket, tapped at the screen. 'Here,' he said. 'Have a look.'

She took the phone and scrolled through the website he had opened. It was a scrappy design peppered with blue links, no logo at the top, just an activist news site done on the cheap. There was a photograph of a man with long hair and slightly boss eyes, squinting into the camera.

Activists expose police spy, read the heading, and then:

'David Hunter' has been an undercover police officer. We are unsure whether he is a serving officer or not. His real name is David Cotmore. Investigations into his identity revealed evidence that he has been a police officer, and a face-to-face confession confirmed this.

Mikey tugged on his cigarette, his arm around Emma, who had her hand on his thigh. Although Emma was still smiling, she looked bored and uncomfortable, like she would rather be anywhere else.

'Remember Dave? Used to help us out, with the van? Then he was with Ali, out in Ireland?'

She shook her head. 'You told me about him. We never met.'

'Okay, well. He was one of them. Ali knew, for a long time. But she's left him now, and blown the whistle. She's talking to the papers. And there are other women, too. It's all about to come out, big-time.'

He flicked the butt out of the window, and there was a little squeal from below, where the smokers were still huddled on the pavement.

'Dan wrote to me today, for the first time in ten years,'

she said. 'I got the message just now. He was saying he wanted to meet.'

'Well, now you know why,' said Mikey. 'He knows the shit is about to hit the fan.'

And then she had to get out, that very second, and away. Away from gatekeeper Mikey and smiling Emma, away from the babble of the smokers and the pretentious fake-art gallery. Away from people who had never had to question anything for a moment in their lives, who took the secure nature of reality for granted. Away from doubts and fears and lies and truth and questions and answers. She stood up and ran full tilt down the stairs, into the gallery space, through the crowd, towards the bright light of outside, elbowing her way through chatting, laughing, drinking people. Normal people, with normal lives. Fuck them all, she thought furiously, barging and shoving them aside, until she reached the front door, where she collided quite hard with Dennis Hopper.

14.

MOLL lived in Sussex now, not far from Henfield. She and Rob, her partner, had bought a small wood, where they ran a forest school. They took kids who were disadvantaged in mainstream education – refugee kids, those with behavioural or learning difficulties. They would come to the forest, where Moll and Rob would let them run around, teach them about the trees and how to make dens.

'Why don't you come and meet her out here?' she had said on the phone. 'The woods are a good place for healing. Remember?'

'I do, Moll,' Skylark replied, remembering the camps, the birdsong, the rustling leaves, the twinkling lights in the trees. Remembering the peace and wild, gentle, happy togetherness of those times. 'I remember.'

'You should see how the kids heal when they come here – they don't want to leave. We make fuck-all money, but we survive, and it's a buzz to see what nature does for them. Yeah, so come, make a day of it, bring Louis, bring whoever. And I'll talk to Ali, see if she'll come and meet you. She's a bit— Well, you'll see. She's not had it easy.'

So they made a day of it. She and Louis went down on the train with Rev and Aoife and Franklin. Moll picked them up from Haywards Heath in her rusty Luton van, with saws and ropes and netting piled in the back, and a nodding Ganesh surrounded by fake flowers on the dashboard. She hadn't changed – still the bovver boots, still the frilly dress over combat trousers. Still the strong, cabbage-hacking arms.

Their land was about half an hour out of town, just outside a small, twee village of red-tiled houses. There was a carved wooden sign by the side of the road, 'The Wild Woods', where Moll pulled off and parked the truck in a clearing. Tall pine trees towered overhead, and the ground was covered with their reddish needles. There was a path leading to a rough canvas shelter, and beside that a sculpture of a stag, made from twisted willow canes.

'Welcome,' said Moll, springing out of the driver's seat. She opened the back door for the boys. 'Go on, then, do what you like.'

Louis and Franklin didn't need to be told twice. They charged out of the car and off into the trees, looking for sticks to make into wands, while Moll led Skylark, Rev and Aoife down the path. 'This is where the kids have lunch,' she said, gesturing to the canvas shelter, which had a row of tables underneath. 'We have twenty at a time, five days a week. Some come every week, others only once. It depends on the school.'

'Where do you live, Moll?' asked Aoife, and Moll pointed upwards. A giant pine had a spiral staircase winding around the trunk, leading up to a tree-house on a hexagonal platform. It had windows looking out over the forest, and a crooked little chimney. A gingerbread house from a fairy-tale.

'Gets a bit wild on windy nights,' she said. 'But we don't mind.'

Further down the path, the woods opened out: no longer dark, shadowy pine, but lighter, dappled beech and ash. In another clearing there was a circle of felled trunks arranged around a stone-lined pit. A small fire was burning in the pit, and a figure was kneeling next to it. As they approached, she stood up and started towards them. Her arms were bone-thin, and her cheeks were hollow. Her hair was buzz-cut, and her bright blue eyes stared out glassily from dark sockets.

'Sky, this is Ali. Ali, meet Sky.' Moll took both women in her strong arms. 'You two have things in common.'

'I've wanted to meet you for so long,' said Ali, grasping Skylark's hand. 'It's funny, I feel like I know you.'

'Oh?' Skylark released herself from the Moll-hug, feeling wary, a little weak. She sat down on one of the trunks. Ali stood with Moll, leaning on Moll, as though she would topple over without her solid body to hold on to.

'I know a lot about you, Skylark,' she said. 'Too much.' Ali closed her eyes, swayed slightly, still holding Moll's hand. 'You keep a vibrator in the top left-hand drawer of your bedroom cabinet,' she said.

Around them, the wind moved through the trees, the sound of sighing. They all stood perfectly still, frozen.

At last Skylark said slowly, 'What the fuck?'

But Ali just swayed, staring, went on in a low monotone: 'On your way back from Spain in February, Customs searched your suitcase, and found two large bottles of sherry and a copy of *Fifty Shades of Grey*.'

'Stop! Shut up!' Skylark cried out, lunging forwards, to tackle her. Moll held her back.

'Just wait, Sky. Listen to her,' she said. 'She's trying to tell you.'

Ali's eyes were open now, her blue eyes burning. She reached out and grasped Skylark's hand, leaned close in to her face. 'I'm not saying this to freak you out. You just have to understand: we have no secrets.' Her hand squeezed more tightly. 'Do you understand? They watch us, and talk about us, among themselves. They know everything, share everything, all the most intimate details. Those most of all.'

'Who do?' hissed Skylark. 'Tell me who they are.'

'The Special Demonstration Squad,' she said. 'The secret police. Nobody in this country even knows it exists.'

'So how do you . . .?'

Now Ali smiled. 'I was living with one of them. For five years. I was an activist once, you see, up north, anti-roads, anti-hunts, animal rights, all that. That's how I met Moll – and how I met Dave. He tried to leave, once – he tried to do just what your Dan did. You know, the psychological breakdown thing. The whole "I just need to get away, to sort my head out, it's not you, it's me."'

Skylark breathed. She spread her roots out into the ground, imagined herself to be a tree, linked by microscopic organisms to all the other trees in the wood. She breathed down into her roots, felt them grow strong and deep.

'But he couldn't do it,' said Ali. 'Came crawling home, begging me to take him back. By that time I had my suspicions – I'd tried to trace him, found that he'd lied about his name and his date of birth. I questioned him and kept on going until he broke down. Admitted that he was a police spy, told me all about the SDS, about the others, about everything. But he said he regretted his choices, that underneath it all he was a world-changer at heart. That he wanted to save the whales, that he always did his recycling. He told me he'd leave the police, if only I'd take him back.'

Ali had knelt down again, was talking to the fire now, to the flames. She was offering her words to it, to the wood, the heat and the light, feeding them in so they could be eaten up and transformed into energy.

'And I really loved him. I wanted to believe he was that person, the one I had got to know. So I took him back. And I waited for him to leave the police. But he never did.'

The fire crackled; sparks flew up into the air. Somewhere, far away, in a different world, Louis and Franklin were casting spells with sticks. Far above them, a light rain had begun to fall on the leaves.

'Instead, he tried to make me leave *my* life,' said Ali. 'He told me to cut off contact with the movement, with my friends. He made us move house, into this horrible gated community where I didn't know anyone. He made me change my name. All the time, he told me we were in danger, that they knew everything about everyone, and that if I told anyone who he really was, he'd be made to pay.'

Moll held her, stroked her back, her shorn hair. Breathed softly, soothingly, into her ear.

'He told me all about you, everything,' said Ali. 'He told me he felt sorry for you, after you had the baby, when you were on your own. Said that Dan had left you high and dry. That he would never have done that, because love and family were more important than work. Whenever I complained, tried to hold him to his word, to make him leave his job, he would remind me that at least he hadn't left me, hadn't done what Dan had.'

They sat in silence for a moment. Tiny pinpricks of rain filtered through the canopy of leaves overhead and landed on them, on their hair, their eyelashes.

'But the truth is, I wish he had,' said Ali. 'Staying was the worst thing he did.'

Leaving the women talking, Rev stalked off into the woods to find Louis and Franklin.

'Wingardium leviosa!'

Following the sound of spells, he found the two boys in a small clearing, and hung back to watch them.

Louis was bossing his younger friend about, as usual. 'You have to take ten steps before we can start the duel, Frankie,' he instructed. 'Close your eyes.'

Franklin was two years younger than Louis, with his mother's red hair and wiry body. He obediently did as he was told, clutching his wand. He didn't have the attention span for Harry Potter books, but he had that younger-brother keenness to do whatever Louis said. On the count of three, they opened their eyes and faced one another.

'Avada kedavara!' cried Louis, unleashing the killing curse.

Franklin howled with rage. 'You can't use that one, Louis! It's too evil!'

'It's okay,' Louis reassured him. 'You can use it against me now. OK, I'm ready – go on.'

Rev stepped into their line of sight. 'Sorry to break up the duel, boys,' he said. 'Franklin, I just wanted to have a word with Louis. Why don't you go and find Mum for a bit?'

Franklin scampered off, waving his wand at a couple of evil-looking tree roots as he went. Rev carefully spread his full-length red PVC jacket on the soft forest floor. 'Come and sit down with me for a minute, Lou,' he said. Louis did so, scratching at his arms absent-mindedly. Rev tenderly stopped him, then lifted his hand to look at the scratches, raised red welts along the inside of Louis's elbow. 'That looks sore.'

Louis shrugged. 'I can't help it.'

'Bodies have a mind of their own, don't they? I realised that when all my hair fell out.'

Louis stroked Rev's bald head. 'I can't imagine you with hair.'

'And I can't imagine you without scratches.'

He put his arm around Louis, who snuggled in closer, rested his head in the crook of Rev's skinny arm. 'Louis, I know you've been wanting to find your dad. I really admire you for that. It's very brave.'

'Not really. It's just normal, isn't it?' Louis picked up his wand, aimed it across the clearing. 'To want to know your family?'

'Of course it is.' Rev stroked his hair. The trees moving around them seemed to be bending over them attentively, protecting them, letting only the finest droplets of rain fall onto their heads and clothes. Beneath them, the forest floor cushioned and supported them, holding them firm. 'I just think it's important to remember,' said Rev, 'who your family really are. They are the people around you, the people who support you through your life. The people who care about you, day in, day out, even when things are not easy.'

Louis was scratching again; again Rev gently stopped him, covered the sore bit with his own hand. 'What I think is, you already know your family, Lou. You've got Mum, Granny, Aoife, Frankie and me. That's a family, isn't it? And it's even better than a normal family, because we've chosen each other. Most families don't get to do that. Can I tell you a secret?' Louis nodded. 'A lot of families don't even like each other very much.'

Around them the trees hushed and swayed. In Rev's arms, the ten-year-old body got heavier, fuller. For a moment, he wondered whether Louis had fallen asleep. But after a moment he said, 'Thank you, Uncle Revvy.'

'Any time, big man.' Rev held him, rested his chin against the dark hair. 'Shall we go and find Mum?'

Louis nodded, and they walked back towards the fire, hand in hand.

'We thought it was just a side issue, didn't we?' Moll was saying. 'That we had to change the world first, and then start thinking about what was wrong between men and women.'

'It wasn't a side issue,' said Aoife.

'It was *the* issue,' said Moll. 'It's all part of the same problem: that we humans have turned against ourselves, against the things that nurture us. We are self-harming, every single day. Poisoning the earth, cutting down trees. Poisoning our own relationships.'

'The people who came up with all this thought that our lives had no value, our bodies had no value, our feelings had no value,' said Aoife. 'We were just tools they could use to get power for themselves.'

'They thought that love didn't matter,' said Moll, looking at Louis, at Franklin, 'when love – real love, which is truth and justice – is the only thing that ever really matters.'

'And now we have to make sure people know what kind of state they're living in,' said Ali, her blue eyes bright. 'This can't have happened to me, to us, for no reason. We always wanted to change the world, didn't we? Perhaps we didn't think it would be this hard, this painful. But change always hurts, it has to.'

Skylark couldn't listen any more. Couldn't try to process what Ali was saying, make sense of her pain, transform it into something improving or meaningful. It was what it was, and her body had had enough.

She laid herself on the soft damp leaves, breathed in their earthy mineral smell. Closed her eyes and sank into the golden space inside: the ancient forests, the standing stones, the inner landscape that somehow remained untouched, no matter what.

One day her body would break down into elements, into water and earth and fire; it would be absorbed by the trees, new leaves would unfurl, and then die, and fall, fall, fall.

The women went on talking; she couldn't hear them any more. She was listening, with her whole body, to something deeper. Could she sense it, the energy moving through the earth? Could she join it, flowing unstoppably onwards, fulfilling the deepest wish of the universe?

Perhaps she could.

Sounds, memories, gathered around the peripheries of her mind, circling like birds, coming in close and then fading away.

Oh-oh, witch-ee-chaio, o-hiyo. Oh-oh, witch-ee-chaio, o-hiyo.

What are you doing? Funny girl.

Lilian, Lilian, are you there? The line is terrible.

What we're talking about here is a total lack of respect for the law.

I love you, both of you. Believe it or not.

Fragments, particles of her life gathered and spun into spirals, into fractals. They broke down into elements, into water, earth and fire. Far above, the blank white sky stared down, big enough to see everything, to contain it, to absorb it all.

Epilogue

A coffee shop. Two people, a man and a woman, are sitting opposite each other at a Formica table. Both have their hands wrapped protectively around their cups. She has curly blonde hair with one grey streak in it. He is tall, thick-set, his dark hair sprinkled with grey and a dimple on his chin.

Let's start right from the beginning, please. What is your name?

Dan. My name is Dan.

Daniel Keller.

That's correct. Yes. Daniel Keller.

Still work for the police?

I do. Counter-terrorism. But I'm on long-term sick.

And – let's see – you're married.

Was. I was married. We split up long ago. Just

after I met you. After that day we went rowing on the canal. Remember?

Yes, I remember. [PAUSE] Your wife – your ex – is Adrienne.

Yes.

And you have two kids?

Not really kids any more. They're sixteen. Twins. Pedro and Anna.

Pedro and Anna. So they were only very tiny when we first met.

They were six months old when I went into the field.

You didn't see them much.

No. Not when they were babies. I was under-cover until they were five. And by the time my tour ended Adrienne and I had been split up for years so – *he shrugs* – it hasn't been easy, for any of us. But I've done what I can. To be a father to them.

And Louis?

Louis.

You're his father, too.

Silence.

He wants to meet you.

Good. I'd love to meet him. I've been keeping an eye, from afar.

Please. Don't say that.

Why?

Don't you understand? It's horrible, sinister to think that you've been watching us all this time, watching and judging, from a distance, like . . . like you were God.

Hardly! I just wanted to keep an eye. Know that you were okay.

But isn't that how you see yourself? Like a special being, with more knowledge and power than everyone else? With the power to decide who knows what, and when, and why? Wasn't that why you did it – the power?

Why I did what?

Everything. Took your position in the watch-tower, up high, on the border between two worlds. Watching one side, watching the other. Taking what you could from both.

That's not how it was.

No?

Believe it or not, I was trying to do the right thing. To keep the peace.

Ha!

She rocks back, casting her eyes up.

Just explain, please, how having a child with someone while lying to that person about everything, everything, even the most basic facts about your identity – just explain to me how that could ever have seemed like keeping the peace.

Sky – please . . .

He catches her hand over the table, to still her, to look into her face.

. . . this is important. It's like I told you, the feelings – they were real. I didn't lie about that. When I said I loved you, you and Louis – I meant it. Everything else may have been lies, but that one thing, that one thing was absolutely true.

Her eyes narrow.

But it can't have been.

It was.

If you love someone, you can't lie to them. If

you love someone, you can't let them have your child, then disappear. That is not love.

Call it what you like. I wanted the best for you. And I left because I thought –

He stops, his voice catches.

– I didn't have any choice. I thought it would be better for you and Louis not to know. I thought you'd be happier without me – that if I stayed, it would cause you more pain.

Her face is in her hands now. It stays there for a long time.

It was a fucked-up situation, Sky. I couldn't change that, and I didn't know what to do. What would you have done?

She looks up, red-eyed. Her face contorts into a snarl of pure, unadulterated fury.

I would never have got myself into that situation.

He snorts.

Well. You weren't in a great situation yourself, when I met you.

What – with Mikey? Listen to me . . .

She grips his hand now, knuckles white, muscles taut, so they are almost arm wrestling.

What you did was a thousand times worse. You made me doubt my own reality. You wormed your way into my life under false pretences. I will never recover from what you did. Every single scrap of joy I have had since I met you has been in spite of what you did.

Hang on, Sky. Just a minute. Look at yourself. Look at your life. You've got a son who loves you, a boss who values you, Rev who supports you, so many friends and family who know what a unique and . . . precious person you are. If anyone's life was ruined . . .

She rolls her eyes and shakes her head.

Oh, please, please, tell me you're not going to ask me to pity you.

He interrupts, his voice ragged.

I loved you and Louis, and I wanted to stay. I told you that, and I meant it. It was a terrible situation, and of course –

He wipes his face on the back of his hand.

– of course I've spent the last ten years wondering how I got myself into that. What

I could have done differently . . . but you know what? It's reality. We can't change it. And if I'd never taken that job, I would have lived my whole life in a small world. I never would have met you, never would have been involved in the world-changing movement, never would have had my eyes opened to all that possibility . . .

Oh, yes – *she oozes sark* – you made sure you got a lot out of it.

And you? Did you get anything out of it?

His eyes have hardened.

You got Louis.

She doesn't reply to this, just chokes, the tears streaming down her face.

I mean it, I'd really love to meet him. He looks like a great kid.

He is. Fuck knows how. He's a miracle.

A long pause.

Is that a possibility? For us to meet?

She fishes a tissue out of her bag, wipes her face.

I don't know. I need to think about what's best for him.

He was the one who wanted to look for me, wasn't he?

She looks up sharply.

How do you . . . *she begins, and then,* I keep getting surprised, that you know stuff! But of course you do. That's your business.

She stands up. He stays sitting, looking up at her. She feels her power, with his eyes on her like that, begging her for forgiveness, to relieve his pain. But this isn't a power she has, or wants. What she wants is to rub him out, to change the past, to stop him ever having existed.

Please think about it, Sky. This is a chance – a chance to make things better. A chance for Louis to have a father. I'm just a human, you know. A messed-up human. I've made mistakes, too many, but deep down, there is something in me that is good.

She looks at him, coolly now, with her clear blue-green eyes. Somewhere in them is still the echo of birds singing, of wind moving through leaves. She has lost a lot of things, but she has never lost that.

I'll think about it.

The woman stands up, walks briskly across the café and

out of the door, emerges blinking into the mundane reality of a Monday morning on City Road.

The man watches her leave, finishes his coffee, then gestures for the bill.

Dedication

To all the people, men and women, whose lives have been affected by the relationships between activists and undercover police officers (and there have been many of you, over decades): as an author I want to acknowledge my feelings of responsibility towards you and your experiences. Working on this book, I tried to meet that responsibility by extending my imagination towards you as fully as I could. In the end, though, this is very much *my* story. It could only ever have been a product of my own psyche and life experiences. As such, it does not aspire to represent you, or anyone other than the characters in my mind.

To the women whose persistence, resilience and determination brought these policing practices to light: I hope that readers of this book will support your campaign for answers, for accountability, and for police reform: https://policespiesoutoflives.org.uk/

Sources

Paul Lewis and Rob Evans's excellent book *Undercover: The True Story of Britain's Secret Police*, based on their reporting of the spy-cops scandal for the *Guardian*, provided the factual underpinning for this project.

The Special Demonstration Squad's *Tradecraft Manual* was published (with redactions) by the Undercover Policing Inquiry in March 2018. It can be read in full here: https://www.ucpi.org.uk/publications/special-demonstration-squad-tradecraft-manual-and-related-documents/

In the recurring phrase 'a flaming arrow of hope', and the chapter on Shepherd's Bush in general, I acknowledge a debt to 'Charlie Fourier', whose vivid account of the 1996 M41 Reclaim the Streets protest can be found here: https://pasttenseblog.wordpress.com/2020/07/13/today-in-london-festive-history-1996-reclaim-the-streets-re-wild-the-m41-motorway-shepherds-bush/

A debt also to Jay Griffiths, whose short but perfectly

formed book *Anarchipelago* gave me Skylark's sentiment 'Father, was where she got it from'. *Anarchipelago*, and *Copse: The Cartoon Book of Tree Protesting* by Kate Evans evoked for me the spirit of 1990s party/protest culture.

Apologies to Manu Chao for taking a liberty with his beautiful earth song, which I turned into a Sky song. It is in fact called 'Por El Suelo'.

Rave tunes quoted in the text are: 'You Gotta Believe' by Fierce Ruling Diva, 'Forward the Revolution' by Spiral Tribe and 'Their Law' by the Prodigy.

I spent much of the 2020 lockdown watching grainy YouTube footage of 1990s parties and protests, which not only helped me with the book, but also reminded me of what freedom felt like. Whoever was sober enough to film these events, and nostalgic enough to put them online: thank you.

Thank you also to: Arts Council England for a project grant, without which this book would never have been completed. ACE is a beacon of light for artists – particularly as, at the time of writing, our government is advising us to 'retrain'. Patrick Morris, for invaluable support with my grant application. Triratna Buddhist order, particularly the *sangha* at the Brighton Buddhist Centre, and most particularly to Dharmakara, and also to the Awakened Heart Sangha, for support, inspiration and regular much-needed affirmation of the value of creativity and the imagination. Kerry Glencorse, Davy Starkey and Charlie Rolfe, for early readings and advice, and to Hannah Black, for asking all the right questions. Johann Hari for the much-needed pep talk. Hazel Orme for her perceptive copy-editing. JC for opening the locker. Paul Rose, for 1990s rave nerdery. Reclaim The Streets, for planting a seed in my mind, all those years ago. People and things that gave me space: Sue Gee, the Gladstone Library, and Janis (my van, my lockdown sanctuary).

And last, but always most of all, to my beloved family: Mandy, Richard and Jessica O'Keeffe, Jonny Morris, and my boys, Stanley and Bowen.